Theories and Modes of
Climate-Smart Agriculture

Theories and Modes of Climate-Smart Agriculture

Bo Li
Ministry of Agriculture and Rural Affairs, China

Xiaogang Yin
China Agricultural University, China

Quanhui Wang
Ministry of Agriculture and Rural Affairs, China

Fu Chen
China Agricultural University, China

Wejin Zhang
Chinese Academy of Agricultural Sciences, China

中国农业出版社
CHINA AGRICULTURE PRESS

 World Scientific

Published by

World Scientific Publishing Co. Pte. Ltd.

5 Toh Tuck Link, Singapore 596224

USA office: 27 Warren Street, Suite 401-402, Hackensack, NJ 07601

UK office: 57 Shelton Street, Covent Garden, London WC2H 9HE

Library of Congress Cataloging-in-Publication Data
Names: Li, Bo (Agronomist), author. | Yin, Xiaogang, author. | Wang, Quanhui, author. |
    Chen, Fu (Agricultural scientist), author. | Zhang, Wejin, author.
Title: Theories and modes of climate-smart agriculture / Bo Li, Ministry of Agriculture and
    Rural Affairs, China, Xiaogang Yin, China Agricultural University, China, Quanhui Wang,
    Ministry of Agriculture and Rural Affairs, China, Fu Chen, China Agricultural University, China,
    and Wejin Zhang, Chinese Academy of Agricultural Sciences, China.
Other titles: Climate-smart agriculture
Description: Hackensack, NJ : World Scientific Pub. Co., [2024] |
    Includes bibliographical references and index.
Identifiers: LCCN 2023040943 | ISBN 9789811283550 (hardcover) |
    ISBN 9789811283567 (ebook) | ISBN 9789811283574 (ebook other)
Subjects: LCSH: Crops and climate--China. | Sustainable agriculture--China. |
    Vegetation and climate--China.
Classification: LCC S600.64.C6 L48 2024 | DDC 630.951--dc23/eng/20240105
LC record available at https://lccn.loc.gov/2023040943

British Library Cataloguing-in-Publication Data
A catalogue record for this book is available from the British Library.

For any available supplementary material, please visit
https://www.worldscientific.com/worldscibooks/10.1142/13601#t=suppl

Desk Editors: Balasubramanian Shanmugam/Steven Patt

Typeset by Stallion Press
Email: enquiries@stallionpress.com

Printed in Singapore

Preface

Each agricultural development mode has a specific significance in each time period, and in different development modes, the main challenges and opportunities faced by the agricultural development stage they are in are explored. In the context of climate change, how to coordinate the relationship between reducing greenhouse gas emissions and ensuring food security, so as to achieve the three goals of mitigating climate change, improving agricultural productivity, and increasing farmer's income to achieve the "triple win" become the major theoretical and technical challenge faced by the world in the 21st century. Under the active advocacy of the Food and Agriculture Organization of the United Nations (FAO), climate-smart agriculture is becoming a new development model for global agriculture in response to climate change.

In order to ensure national food security, actively respond to climate change, and promote the development of green and low-carbon agriculture, with the support of Global Environment Facility (GEF), the Ministry of Agriculture and Rural Affairs (former Ministry of Agriculture, renamed by April 3, 2018), and the World Bank jointly implemented China's first climate-smart agriculture project: Climate Smart Staple Food Crop Production Project (2014–2020).

Over the past five years, the project has successfully transformed the internationally advanced climate-smart agriculture concept into the success practice of China's agricultural in response to climate change, explored and established a variety of resource-efficient, economically rational, carbon-sequestrating, and emission-reducing food production technology models, and achieved the "triple wins": mitigating climate change, improving agricultural productivity, and increasing farmers' income. It has developed a new path of China's agricultural green development, contributing Chinese experience and wisdom to sustainable development of global agriculture.

The "Climate Smart Staple Food Crop Production Project" continuously improved the project design by inviting internationally renowned experts to participate in the design, seminars, on-site guidance, on-site inspections, and opinion exchanges, which reflected the designed concept of "global vision" and "China's conditions". Through the joint participation of managers, experts, entrepreneurs, and farmers, the project has achieved a good environmental evaluation and social assessment effect with "participation of farmers and women". Based on the successful implementation of the project and valuable experience gained, we have compiled the "The Series of Climate Smart Agriculture" (12 volumes) to further summarize and improve the theoretical system, measurement methods, technical models, and development strategies of climate-smart agriculture, tell the story of China's climate-smart agriculture, promote the wider dissemination, and application of climate-smart agricultural concepts and good practices in China and the world.

As the epitome of China's climate-smart agriculture practices, "The Series of Climate Smart Agriculture" features theoretical, practical, and strategic aspects of the project. A variety of flexible expression forms and contents, including theoretical research, strategic recommendations, methodological guidelines, case studies, technical manuals, and brochures. Targeted at a wide readership, it can be used as a reference for decision-making by agricultural and rural department managers, as well as by agricultural technology extension personnel to guide farmers to carry out front-line practice. It can also be used as a reference book for teaching in agricultural colleges and universities.

Climate-smart agriculture has just started its development and practice in China, and related theories and technical models need to be further systemized and systematized. More related research fields need to be further expanded, especially the integrated management technology of climate-smart agriculture and ecological landscape-based regional management model need to be further explored. Due to limitations of time, energy, and research level of the editors, there are still some deficiencies in this series. We hope to use these series of books to attract more people and look forward to feedback and suggestions to jointly promote the development of China's climate-smart agriculture, ensure China's food security, achieve China's 2060 carbon-neutral goals, and contribute to the strategic transformation of China's agricultural production mode.

About the Chief Editors

Bo Li is the Deputy Director General of the Department of Science, Technology and Education, Ministry of Agriculture and Rural Affairs. He is in charge of agricultural resources environmental protection and rural energy ecology and has participated in the implementation of international environmental conventions, such as the Global Alliance for Agricultural Greenhouse Gas Research, the United Nations Framework Convention on Climate Change, the United Nations Convention on Biological Diversity, and the Montreal Protocol on Substances that Deplete the Ozone Layer on behalf of China.

Xiaogang Yin has been working as an Associate Professor in the College of Agronomy and Biotechnology, China Agricultural University since January 2018. Xiaogang Yin received his PhD degree in agronomy from China Agricultural University in 2015, and afterward he started post-doc research on C/N cycling in crop rotation-based crop models and long-term experiments with the support of MACRUR project and Marie Curie AgreenSkills+ Fellowship in Aarhus University and INRA, respectively. Xiaogang is experienced in crop modeling, C/N cycling in cropping systems and climate change; his current research interests include (1) sustainable development of cereal-legume rotations in China, (2) C & N cycling in cropping systems based on crop modeling and experiments and (3) climate-smart agriculture. Xiaogang has published more than 40 peer-reviewed scientific papers, carried out 5 national projects, and won 2 first prizes in science and technology progress at provincial and ministerial levels.

Quanhui Wang is a Researcher and the Chief Expert of the Rural Energy and Environment Agency, Ministry of Agriculture and Rural Affairs, who has long been engaged in international exchanges and demonstration introduction of international projects in the field of sustainable agriculture, promoted multilateral and bilateral international cooperation and exchanges on agricultural resource in China, designed and led several international projects including PRC-GEF Partnership Program on Sustainable Agricultural Development and Market Transformation of Energy Efficient Bricks and Rural Buildings Project, Enabling Zero Carbon Energy in Rural Towns and Villages in China Project, and the ADB project Study of Clean Energy Supply for the Rural Areas in the Greater Beijing-Tianjin-Hebei Region, Innovative transformation of China's food production systems and agro-ecological landscapes toward sustainability, etc. and dedicated to promoting green development and ecological transformation of agriculture in China.

Fu Chen got his PhD from Beijing Agricultural University in 1991 and since then has been working in the College of Agronomy & Biotechnology at China Agricultural University. He is the Chairmen of the Chinese Society of Farming Systems Research, the Director of the Key Laboratory on Farming System of the Chinese Agricultural Society and a member of the National Carbon Neutral Science and Technology Expert Committee and enjoys the special allowance of The State Council. His research work has focused on crop cultivation technology improvement, farming system optimization, conservation tillage and climate change adaptation. He is also the Chief Technical Advisor of China

Climate Smart Staple Crop Production Project (2015–2020) supported by GEF and World Bank. As the team leader, Prof. Fu Chen has completed more than 10 national research projects. He has published more than 10 monographs and textbooks, published more than 200 academic papers, and won 5 national and provincial scientific and technological achievement awards.

Weijian Zhang is the Chief Scientist of Agro-ecology and Farming System, Chinese Academy of Agricultural Sciences. He graduated from Nanjing Agricultural University in 1999 with a doctor's degree in crop cultivation and farming system. For many years, he has been engaged in teaching and scientific research in the field of farmland ecology and farming system, focusing on the response and adaptation of farmland ecosystem to global change, farmland ecological health and farming system, agricultural system evolution and regional agricultural development. In recent years, he has presided over 973 national projects, National Natural Science Foundation of China, National Science and Technology support plan, doctoral program of the Ministry of Education and Jiangsu Natural Science Foundation. He has won two second and third prizes in science and technology progress at provincial and ministerial levels, and relevant research has been published in influential journals. He also edited and participated in the compilation of several teaching materials in important journals.

About the Associate Editors

 **Xin Zhao** teaches agronomy at the College of Agronomy and Biotechnology, China Agricultural University (CAU). He received his PhD and BASc in Agronomy from CAU. His research interests focus on conducting meta-analysis in agronomy, establishing conservation agriculture in China and soil organic carbon sequestration and greenhouse gas emission from agricultural practices. In addition to his current position at CAU, Xin has served as the editorial board member for Land, Farming System, and Soil and Environmental Health. Recently, Xin aims to reveal the mechanisms of crop residue retention and no-till to sequestrate soil carbon and reduce greenhouse gas emissions with site-specific field experiments and meta-analysis. He has published more than 40 academic papers in journals including *Global Change Biology, Advances in Agronomy*, the *Journal of Environmental Management*, and the *Science of the Total Environment*.

 **Haotian Chen** currently serves as a Postdoctoral Researcher at the French National Research Institute for Agriculture, Food, and the Environment (INRAE). He holds a PhD in environmental science from Université Paris-Saclay in France and an MPhil from China Agricultural University in China. Dr Chen's research is dedicated to advancing sustainable practices, with a specific focus on optimizing organic waste utilization, mitigating greenhouse

gas emissions during composting and understanding soil nutrient cycling dynamics. His scholarly contributions have been featured in academic journals, including publications in *Soil and Tillage Research*, the *Journal of Environmental Management*, and others.

Dahai Guan got a Doctorate degree in Agronomy from China Agricultural University in 2014. Since then, he has successively worked in the Rural Energy and Environment Agency, the Ministry of Agriculture and Rural Affairs (MARA) and the Department of International Cooperation, MARA. He was a negotiator of the Chinese delegation for the United Nations Framework Convention on Climate Change (UNFCCC) and a member of the "Technical system of modern agricultural industry" for rape crops. His research interests include climate change adaptation and mitigation and he dedicates himself to the research and replication of the practice of climate-smart agriculture. More than 20 publications and papers have been published in the last five years.

Qingmei Wang received her PhD from the Chinese Academy of Agricultural Sciences (CAAS) in 2020 and is currently conducting postdoctoral research at the Institute of Agricultural Resources and Regional Planning, CAAS. Her research mainly focuses on nitrogen cycling in the cropping and livestock systems and the associated environmental impacts due to losses of reactive nitrogen. Supported by the China Scholarship Council, she even visited the University of Melbourne as a joint PhD student from 2018 to 2020 and since then has developed good international collaboration with the University of Melbourne (Australia), the University of Alberta (Canada), etc. She has been involved in several national and international research projects and published more than 10 peer-reviewed papers in international journals by far. Prior to her PhD, she worked as a project assistant in REEA, MARA from 2014 to 2018 and participated in the implementation and management of the China Climate Smart Staple Crop Production Project (2015–2020) funded by GEF and the World Bank.

List of Contributors

Fu Chen *College of Agronomy and Biotechnology, China Agricultural University (CAU)*

Haotian Chen *French National Research Institute for Agriculture, Food, and the Environment (INRAE)*

Dahai Guan *Rural Energy and Environment Agency, Ministry of Agriculture and Rural Affairs*

Wenhai Huang *College of Agronomy and Biotechnology, China Agricultural University (CAU)*

Bo Li *Department of Science, Technology and Education, Ministry of Agriculture and Rural Affairs*

Zhao Liu *Rural Energy and Environment Agency, Ministry of Agriculture and Rural Affairs*

Wensheng Liu *College of Agronomy and Biotechnology, China Agricultural University (CAU)*

Shanheng Shi *College of Agronomy and Biotechnology, China Agricultural University (CAU)*

Meifang Wang *Anhui Provincial Department of Agriculture and Rural Affairs*

Qingmei Wang *Institute of Agricultural Resources and Regional Planning, CAAS*

Quanhui Wang *Rural Energy and Environment Agency, Ministry of Agriculture and Rural Affairs*

Xiaohui Wang *College of Agronomy and Biotechnology, China Agricultural University (CAU)*

Lin Xue *Rural Energy and Environment Agency, Ministry of Agriculture and Rural Affairs*

Ying Yang *College of Agronomy and Biotechnology, China Agricultural University (CAU)*

Xiaogang Yin *College of Agronomy and Biotechnology, China Agricultural University (CAU)*

Li Zhang *College of Agronomy and Biotechnology, China Agricultural University (CAU)*

Weijian Zhang *Agroecology and Farming System, Chinese Academy of Agricultural Sciences*

Yanping Zhang *Research Associate of Food and Natural Resources Program, China, World Resources Institute (WRI)*

Xin Zhao *College of Agronomy and Biotechnology, China Agricultural University (CAU)*

Jun Zou *College of Agronomy and Biotechnology, China Agricultural University (CAU)*

Contents

Chapter 1
Introduction

Abstract

The introduction summarizes the relationship between global climate change and agricultural production, defines the background of climate-smart agriculture (CSA), outlines the theoretical framework of CSA, clarifies the objectives of CSA, and concludes with the preliminary practical effects of CSA in China. Specifically, CSA is a new approach to guide the transformation of agricultural production modes, aiming to solve the common issues of food security and climate change. CSA has three core objectives of: (1) improving the ability of agricultural production to adapt to climate change; (2) reducing or removing agricultural greenhouse gas emissions; (3) sustainably increasing agricultural productivity and incomes, which is the main development direction of global agricultural production to cope with climate change. CSA basically coincides with China's agricultural transformation and development goals, and it is an important way to promote the green and sustainable development of China's agriculture. The successful implementation of CSA crop production projects in China has promoted the theoretical development and practical application of CSA, providing scientific support for China's agricultural transformation.

1.1 The Proposal of Climate-Smart Agriculture

Ensuring human food security and mitigating the impact of climate change are dual challenges facing the global society today. It is projected that the global population will reach 9 billion in 2050, with 2.4 billion of the newly added population living in underdeveloped countries, and global agriculture will need to increase productivity by at least 60% to feed all humanity (FAO, 2009). Therefore, maintaining sustained and steady growth in agricultural productivity is the inevitable way to reduce poverty and ensure food security. However, climate change, characterized by rising atmospheric temperature and frequent occurrence of extreme weather events, has already impacted global agriculture in various ways (IPCC, 2014). Relevant studies have shown that climate change has led to a 3.8% and 5.5% decline in average global maize and wheat yields, respectively. Provided that the atmospheric temperature continues to rise in the future, more severe crop yield reductions will occur. Especially in low-latitude regions, even a small increase in temperature (1–2°C) can significantly affect the productivity of plantation and graziery (Battisti *et al.*, 2009; Lobell *et al.*, 2011), which threatens global food security. The increasing concentration of greenhouse gases, such as CH_4, N_2O, and CO_2 in the atmosphere caused by human interference, is the main driving factor of climate change, while agricultural production and related land-use changes (mainly deforestation) have led to significant greenhouse gas emissions, accounting for about 30% of the total global greenhouse gas emissions caused by human activities. In addition, animal husbandry accounts for 18% of global greenhouse gas emissions (IPCC, 2013). The traditional agricultural development model, which is based on massive investment in resources, led to the degradation of biodiversity and ecosystem services, as well as the pollution of soil and water (Lipper *et al.*, 2014). In recent decades, many countries and institutions have issued a series of policies to deal with climate change, increase yield and grower incomes, and promote agricultural sustainable development, but most of these policies focused on certain aspects and a part of climate change policy covered sustainable agricultural production. Hence, effective methods are needed to improve agricultural production and ultimately achieve the mandated goals of ensuring food security and climate change mitigation. Based on these circumstances, the Food and Agriculture Organization of the United Nations (FAO) pioneered the concept of climate-smart agriculture (CSA) on The Hague Conference on Agriculture, Food Security and Climate Change in 2010. As a comprehensive policy, CSA emphasizes the possibility of increasing agricultural productivity, contributing to poverty eradication and making agriculture more adaptable to climate change. In addition, it is effective to reduce greenhouse gas emissions and increase the carbon fixation capacity of crop production systems (FAO, 2013).

In 2013, FAO published the book *Climate-Smart Agriculture Sourcebook*, which elaborated in detail the development concepts, research methods, and related cases of CSA from 18 aspects: background and goal, land management, water resources management, soil management, rational agricultural management, protection and sustainable utilization of agricultural and food genetic resources, crop production system, animal husbandry, forestry, fishery, sustainable food chain, community organizations, mainstreaming in national policies, financing, disaster risk reduction, agricultural insurance, agricultural capacity building, assessment, and evaluation. It lays a theoretical foundation for the development and application of CSA in the global arena (FAO, 2013).

1.2 Goals and Objectives of CSA

1.2.1 *Goals of CSA*

CSA is an integrated system of technologies, policies, and investments to achieve sustainable agricultural development and food security under climate change; it aims to respond the challenges of food security and climate change better by integrating three important aspects of sustainable development: economy, society, and environment. CSA includes three essential objectives:

(1) Improve the adaptability of agricultural production against climate change;
(2) Minimize agricultural greenhouse gas emissions as possible;
(3) Increase agricultural productivity and grower incomes sustainably.

It is the main developing direction for global agricultural production to respond climate change (FAO, 2013).

Enhancing the adaptability and resilience of agriculture to respond to climate change is one of the goals of CSA: The core is to build the resilience of food production systems and food producers against climate change and enable them to adapt to long-term climate change. The increasing impact of climate change on agricultural production has severely affected food availability and food shortages could lead to soaring of prices (FAO, 2018). Therefore, establishing the resilience of the agricultural system to climate change is essential for guaranteeing food security, which could help to ensure food supply and intra- and inter-annual stability in food production. The main strategies to achieve the goal of enhancing the adaptability and resilience of agriculture to climate change include: diversifying production systems, planning production activities to reduce risks and sensitivity, adapting changing conditions, and managing agroecosystems, ecosystem services, and biodiversity.

Minimizing greenhouse gas (GHG) emissions is an important goal of CSA: CSA emphasizes on the reduction of GHG emissions from food production in terms of absolute emissions and emission intensities. The Intergovernmental Panel on Climate Change (IPCC) found that limiting global warming to 2.0°C above pre-industrial level will help achieve sustainable development goals; it will reduce the number of people exposed to climate risks and vulnerable to poverty and increase the possibility for poverty eradication and reducing inequalities. Climate change mitigation measures in agriculture are important to respond to climate change. The main strategies for achieving CSA and minimizing GHG emissions include improving resource utilization efficiency, enhancing the carbon fixation capacity of agroecosystem, and replacing fossil fuel-based energy by renewable energy.

Increasing agricultural productivity and growers' incomes sustainably is the essential goal of CSA: CSA should integrate economic, social, and environmental factors to ensure food security. Sustainably increasing agricultural productivity and growers' incomes mainly guarantee food security in two ways: increased productivity increases the food supply for household and local markets; improved incomes further secure the food supply. The main strategies based on CSA to sustainably increase agricultural productivity and grower incomes include improve resource utilization efficiency; diversify production systems; and manage agroecosystems, ecosystem services, and biodiversity.

1.2.2 *Objectives of CSA*

CSA needs to comprehensively accord with local climate, natural environment, market demand, and economic and cultural situations. Therefore, the degree, directness, and scope of the impact of climate change on the agricultural system should be integrated into CSA planning (Celeridad, 2018). The practical results of a large number of CSA projects show that achieving the multiple goals of CSA requires an integrated approach tailored to local situations, and the synergy between the agricultural sector and other sectors is very important to optimize the use of natural resources and increase the value of ecosystem services. CSA is not a settled specific agricultural technology or method that can be applied universally. It needs to be adapted to different development goals. The main objectives of developing CSA include the following:

(1) Identifying integrated solutions to the interrelated challenges of food security, social development, and climate change mitigation that can generate multidimensional synergies and effectively reduce the overall benefit loss;
(2) Tailoring specific plans to context-specific and relevant social, economic, and environmental backgrounds; fully assessing the interactions between departments and the needs of different stakeholders;
(3) Clarifying the obstacles to the application of CSA to farmers and proposing solutions in terms of policies, strategies, actions, and incentive measures that strive to coordinate and create an enabling environment for scaling up through policies, financial investment, and institutional arrangements;
(4) Achieving multiple goals, prioritizing them, and making collective decisions about the benefits and trade-offs;
(5) Giving priority to increasing farmers' incomes by improving knowledge dissemination methods, resource allocation methods, service methods, financial policies, and market allocation methods;
(6) Increasing the resilience of agricultural production to climate change shock, and mitigating greenhouse gas emissions from agricultural production is a potential secondary co-benefit;
(7) Seeking climate-related financial support and use it in conjunction with traditional agricultural investments.

By integrating extension applications, policy formulation, and service agencies, CSA can help address the multiple challenges facing the agricultural food system.

CSA is a new approach to guide the transformation of agricultural production methods, aiming to solve the common problems of food security and climate change. It is closely linked to sustainable intensification, with the goals and guiding principles of sustainable development and green economy, contributing to food security and natural resources (FAO,

2011). CSA is not a new agricultural system or a specific technical guidance. It integrates the three aspects of sustainable development to solve the problems of food security and climate change, helping to achieve the specific goals of sustainable development. CSA addresses urgent global problems at present, highlighting the practicality and necessity of sustainable development. The emphasis of CSA is to improve productivity, increase farmers' incomes, and ensure stability by focusing on crucial aspects related to the security, availability, and stability of food. Diversification of production methods is an effective way to improve system resilience and its efficiency, and it is also the path to achieve a more balanced and nutritious diet. CSA and sustainable intensification of crop production have common goals and principles. Sustainable intensification means that agricultural production should protect and use natural resources more efficiently; compared to it, CSA has a forward-looking dimension and is more attentive to potential future changes and makes necessary preparations for these changes.

CSA pays great attention to the livelihoods and food security of smallholder farmers and emphasizes the use of appropriate methods and technologies to produce, process, and sell agricultural products by improving the management and utilization of natural resources. In order to maximize benefits and minimize losses, CSA not only considers the social, economic, and environmental background, but also evaluates the impact on energy and local resources. The successful implementation of CSA necessarily requires related policy, technical and financial support, and climate change should be considered inside of agricultural development plans to achieve green and sustainable agricultural development. The extension of CSA needs the support of appropriate mechanisms and systems, ensuring extensive participation and policy coordination. In practice, the multiple goals of CSA cannot be achieved once and for all, and related evaluation mechanisms should be established to prioritize different goals according to the actual situation and focus on specific goals and evaluation basis vary with phases.

1.3 CSA Practices in China

CSA basically coincides with China's agricultural transformation and development goals, mainly in the following aspects:

Firstly, it emphasizes that food security and agricultural development cannot be affected by coping behaviors of climate change. It highlights the importance of adaptation and mitigation of climate change to achieve high crop yield, high resource efficiency, and carbon sequestration and emission reduction together, which is consistent with China's agricultural green development goals.

Secondly, it emphasizes policy and system innovation, benefits coordination and win–win cooperation, and the establishment of coordination mechanisms and flexible management systems between departments and industries, which is key support for promoting the transformation and development of China's agriculture.

Finally, it also emphasizes innovative technology research and extension models and builds a set of production models and comprehensive technical systems that are compatible with policies and management systems, which meet the needs to promote institutional technological progress in China's agriculture.

Therefore, according to the concept of CSA development, with the support of the Global Environment Facility and the World Bank (WB), the Chinese Ministry of Agriculture and Rural Development organized and implemented the Climate-smart Staple Food Crop Production Project in 2015, providing theoretical and practical support for the development of CSA in China.

The Climate-smart Staple Food Crop Production Project explores the development path of CSA based on Chinese characteristics and mainly includes four aspects: technology demonstration and application, policy and system innovation, knowledge management, and monitoring and evaluation. The Climate-smart Staple Food Crop Project has established 3,333 hectare of demonstration field in the major grain-producing areas, Huaiyuan County, Anhui Province, and Ye County, Henan Province to demonstrate climate-smart production models for wheat–rice production system and wheat–corn production system, respectively. It aims to explore policies and measures related to carbon sequestration and emission reduction in staple food crop production systems, enhance the environmental protection awareness of project beneficiaries, improve the quality of personnel in the project area and the use and management levels of main agricultural inputs, improve the utilization efficiency of agricultural inputs such as water, fertilizers, pesticides, and the operational efficiency of agricultural machinery for the three major staple crops in China, rice, wheat, and corn, reduce GHG emissions from the crop production system, increase the amount of carbon storage in farmland soil, enhance the adaptability of the crop production system to climate change and achieve a sustainable increase in crop yield and grower incomes. In 2016, the

"Northwest Oasis Climate-smart Agricultural Ecological Efficient Three-dimensional Cultivation Demonstration Project", supported by the Global Environment Facility, was also implemented.

This book, funded by the Climate-smart Staple Food Crop Production Project (P144531), is a systematic review of relevant reports on CSA by FAO, the World Bank, and the Consultative Group on International Agricultural (CGIA) over the past 10 years, and extensively collected domestic and international literature resources related to climate change and its impact on agricultural production and adaptation, agricultural GHG emission reduction, and conservation agriculture in the past 20 years. This book comprehensively and deeply summarized the development status and trends, research methods and development strategies, monitoring, evaluation, and extension of global CSA. And it explains in detail the current status and trends of China's CSA, the development status and trends of international CSA, and it provides scientific support for the development of China's CSA.

Combined with the implementation of the Climate-smart Staple Food Crop Production Project, we systematically sort out the relevant experience and shortcomings obtained since the implementation of the project over the past 5 years from the perspective of practice and exploration of climate-smart wheat–rice production and climate-smart wheat–maize production in China.

Based on these, combined with the current situation, experience, and challenges of domestic CSA development and comprehensive consideration of agricultural green sustainable development policies in China, we put forward policy recommendations and technical systems for the future agricultural development in China and hope to provide scientific support for the realization of green sustainable development of China's agriculture.

The book consists of nine chapters as follows.

Chapter 1: Introduction
Chapter 2: Current Status of International Climate Smart Agriculture Development
Chapter 3: International Climate Smart Agriculture Practice Approach
Chapter 4: Monitoring, Assessment and Extension of Climate-Smart Agriculture Projects
Chapter 5: International Climate-Smart Agriculture Cases
Chapter 6: Background of China's Climate-Smart Agriculture Development

Chapter 7: Practice and Exploration of Wheat–Rice Production with Climate-Smart Agriculture in China

Chapter 8: Practice and Exploration of Wheat–Maize Production with Climate-Smart Agriculture in China

Chapter 9: Summary of China's Climate Smart Agriculture Mode Experience

References

Battisti D S, Naylor R L. 2009. Historical warnings of future food insecurity with unprecedented seasonal heat. *Science*, 323: 240–244.

Celeridad R. 2018. *Contextual and Universal: Scaling Context-specific Climate-smart Agriculture*. World Agroforestry Centre. Nairobi, Kenya.

FAO. 2009. *How to Feed the World in 2050. Issues Brief for the High-level Forum on How to Feed the World in 2050*. Rome: FAO.

FAO. 2011. *Save and Grow: A Policymaker's Guide to the Sustainable Intensification of Smallholder Crop Production*. Rome: FAO.

FAO. 2013. *Climate-smart Agriculture: Sourcebook*. Rome: FAO.

FAO. 2018. *Transforming Food and Agriculture to Achieve SDGs*. Rome: FAO.

IPCC. 2013. Climate Change 2013: The Physical Science Basis. Contribution of Working Group I to the Fifth Assessment Report of the Intergovernmental Panel on Climate Change [Stocker, T.F., D. Qin, G.-K. Plattner, M. Tignor, S.K. Allen, J. Boschung, A. Nauels, Y. Xia, V. Bex and P.M. Midgley (eds.)]. Cambridge University Press, Cambridge, United Kingdom and New York, NY, USA, 1535 pp.

IPCC. 2014. Climate Change 2014: Impacts, Adaptation, and Vulnerability. Part A: Global and Sectoral Aspects. Contribution of Working Group II to the Fifth Assessment Report of the Intergovernmental Panel on Climate Change [Field, C.B., V.R. Barros, D.J. Dokken, K.J. Mach, M.D. Mastrandrea, T.E. Bilir, M. Chatterjee, K.L. Ebi, Y.O. Estrada, R.C. Genova, B. Girma, E.S. Kissel, A.N. Levy, S. MacCracken, P.R. Mastrandrea, and L.L. White (eds.)]. Cambridge University Press, Cambridge, United Kingdom and New York, NY, USA, 1132 pp.

Lipper L, Thornton P, Campbell B M, *et al.* 2014. Climate-smart agriculture for food security. *Nature Climate Change*, 4: 1068–1072.

Lobell D B, Schlenker W, Costa-Roberts J. 2011. Climate trends and global crop production since 1980. *Science*, 333: 616–620.

Chapter 2

Current Status of International Climate-Smart Agriculture Development

Abstract

Over the past century, global climate change, mainly arising out of temperature rise and uneven distribution of rainfall over time and space, has led to frequent occurrences of extreme weather events, such as high temperature, drought, and floods, which have caused uncertainties in food security and carbon sequestration, reducing GHG emissions, stabilizing crop production, and increasing farmers' incomes. How to effectively deal with the challenges caused by climate change is still a global concern. Climate-smart agriculture (CSA) aims to better respond to the challenges of food security and climate change by integrating three important aspects of sustainable development: economy, society, and environment. It is a key approach to implementing the sustainable development of international agriculture and provides new ideas and solutions for mitigating climate change and ensuring food security. International CSA mainly includes three aspects: building climate-smart crop production system, improving systems and policies, and financing and capacity building. International institutions such as the Global Environment Facility (GEF), the World Bank, and the Food and Agriculture Organization of the United Nations (FAO) have led a number of CSA projects in different countries and regions, which promoted the establishment of the International Coalition for CSA and resulted in the development of theories and practices of CSA. Based on the achievements of the past 10 years of CSA in various countries, the experience for future CSA development mainly includes three aspects: (1) formulating the detailed goals and plans of CSA; (2) emphasizing on the integration and application of new materials, new technologies, and new methods in the practice of CSA; and (3) supporting projects with long-term stable funding. Developing countries are facing more severe challenges during the development stage of CSA. They demand favorable national policy environments, available data acquisition and information dissemination channels, and practical subsidy incentive mechanisms and insurance services.

2.1 Climate Change Has Exacerbated the Challenges of Food Security, Carbon Sequestration, Emission Reduction, and Stable Production and Income

(1) Climate change is a key factor affecting agricultural production. The impact of climate change on crop production is a global concern (Lobell *et al.*, 2011; IPCC, 2013; Lipper *et al.*, 2017). Climate change characterized by frequent occurrences of global warming and extreme weather events has become a reality. According to the Fifth Assessment Report from IPCC, global surface temperature has risen by 0.89°C in the past 100 years, and the frequency and intensity of agro-meteorological disasters such as heat stress and drought have increased significantly. The global average rainfall did not change significantly, but extreme droughts and floods have occurred frequently in some areas (IPCC, 2013). Global warming has brought multiple effects on agricultural production (Lobell *et al.*, 2011; Trnka *et al.*, 2014; Ray *et al.*, 2015). Studies have shown that if the daily maximum temperature is higher than 30°C, it will adversely affect corn production. And each 1°C temperature increase could cause 1.7% yield reduction in corn (Schlenker and Roberts, 2009; Lobell *et al.*, 2011, 2013). The frequent occurrence of extreme weather events such as high temperature and drought in the context of global warming has caused serious crop yield reductions of wheat and corn and posed a serious threat to global food security (Webber *et al.*, 2018; Kahiluoto *et al.*, 2019). IPCC predicted that the global temperature will keep rising before 2050 under different emission scenarios to meet the needs of humans and society to survive and develop (Mannet *et al.*, 2009a; IPCC, 2013). This will bring exacerbated risks to crop production in form of high temperature and drought (Trnka *et al.*, 2014; Deryng *et al.*, 2014), leading to more serious crop yield reductions.

Appropriate adaptation strategies have to be devised to deal with the adverse climate effects to maintain the productivity of agricultural systems and even provide new opportunities for improving the sustainability of the system (Moore and Lobell, 2014; Adenle *et al.*, 2015; Burke and Emerick, 2016). Adaptation to climate change is the process of adapting natural or human-made systems to enhance their resilience to the impacts of anticipated climate change, including active and automatic adaptations (Smit and Wandel, 2006; Barrett *et al.*, 2007; Olesen *et al.*, 2011). Active adaptation indicates that the government adjusts relevant policies to make climate change adaptation measures more effective and mainly includes strengthening infrastructure construction, increasing investment in resistance breeding, improving disaster resistance systems and regulations, and providing disaster risk subsidies (Howden *et al.*, 2007). Automatic adaptation means that farmers adopt technical adjustments to cope with climate change, whose technologies mainly include adjusting crop sowing dates, selecting more suitable crop varieties, changing soil cultivation techniques, optimizing water and fertilizer management, and strengthening control of plant diseases and pests (Olesen *et al.*, 2011). The actual impact and climate change are closely related to the characteristics of farmers, mainly because they affect the effectiveness of management measures and the use of adaptation technologies.

(2) Mitigating climate change is an important part of sustainable agricultural development: The increasing concentration of greenhouse gases, such as CH_4, N_2O, and CO_2 in the atmosphere caused by human activities, is the main driving factor for accelerating climate change. Agricultural production and related land use changes, especially deforestation, are important reasons for greenhouse gas emissions, accounting for about 30% of global greenhouse gas emissions; greenhouse gas emissions caused by deforestation and farmland reclamation accounts for about 17% of global emissions (IPCC, 2007b; IPCC, 2013). Agricultural production is not only an important source of CO_2 emissions but also an important source of N_2O and CH_4 emissions, accounting for 60% and 61% of global N_2O and CH_4 emissions, respectively (Celeridad *et al.*, 2018). N_2O emissions are mainly caused by soil activities and fertilizer applications, while CH_4 emissions are mainly due to rice planting and ruminant intestinal digestion

((Mann *et al*., 2009a) IPCC, 2013). With the increase in agricultural inputs, the IPCC predicted that the N_2O and CH_4 emissions caused by agricultural production will increase by 35–60% and 60%, respectively, by 2030 (IPCC, 2007b; Raney *et al*., 2009b). Agricultural greenhouse gas emissions depend on natural processes and agricultural production management, which can be effectively slowed down by reasonable management measures. Agricultural production can alleviate climate change in two ways: reducing emissions per kilogram of grain produced and enhancing the function of soil carbon sinks. The IPCC estimated that in 2030 the global improvement in agricultural technology can achieve a reduction of 5,500–6,000 tons of CO_2 per year, accounting for about 75% of the emissions of the agricultural industry. Developing countries could contribute 70% of the reduction potential, and the Organization for Economic Co-operation and Development (OECD) countries and transitioning countries will respectively contribute 20% and 10% of the emission reduction potential (IPCC, 2007b). The IPCC estimated that 90% of the global agricultural emission reduction potential comes from reasonable soil carbon pool management, including increasing soil carbon sequestration, reducing soil disturbance, improving grazing management, increasing soil organic matter, and restoring degraded land (IPCC, 2007b; FAO, 2009c).

(3) Ensuring food security and farmers' livelihoods are the core of agriculture in responding to climate change: Food security is a focal issue in global communities, with FAO estimating that approximately 820 million people are undernourished and 9.2% of the global population has severe food crisis, necessitating international emergency support (FAO, 2019). Importantly, 60% undernourished people are food producers, and for these poor producers, food is not only a basic need but also the only source of livelihood. FAO projected that the global population will reach 9.1 billion in 2050, with about 2 billion newborn people living in economically disadvantaged countries. At the same time, with the acceleration of urbanization, 70% population will live in urban areas, which will accelerate the growth of animal product consumption (FAO, 2018b). In view of this, FAO estimated that global food production must increase by 70% in 2050 to meet the expected demand for food and forage (FAO, 2018a). In addition, the global climate has deteriorated continually in recent decades, which significantly affected agricultural production and posed a significant challenge to global food security (IPCC, 2014). Therefore, maintaining sustainable and stable growth of agricultural production capacity in the context of climate change is the inevitable way to reduce poverty and ensure food security. The future development of CSA will focus more on ensuring food security and increasing farmer incomes, while improving the resilience and endurance of the agricultural system to respond to climate change and reducing greenhouse gas emissions. This is a key link in the development of CSA.

2.2 CSA: The Key Approach to Sustainable Agricultural Development

The growing global population, escalating food security crisis, and the increasing impact of global warming on agriculture have made the application of CSA more urgent. In view of this, CSA provides globally applicable principles for agricultural food security management and provides scientific support for relevant policy recommendations of multilateral organizations such as the United Nations Food and Agriculture Organization (Roy *et al.*, 2018). CSA addresses the relevant limitations of the international climate policy field and is based on an in-depth understanding of the important role of agriculture in food security and the potential synergy between agriculture's adaptation to climate change and reduction of greenhouse gas emissions. CSA is committed to achieving coordinated and sustainable increases in agricultural productivity and farmer incomes, building agricultural resilience and adapting to climate change, as well as reducing and eliminating agricultural greenhouse gas emissions. Furthermore, CSA needs to consider the entire sustainability of its achievements, aiming to get truly positive and lasting results during the fight against hunger and climate change. The "Transforming Our World: The 2030 Agenda for Sustainable Development" clarified 17 Sustainable Development Goals and comprehensively explained the objectives of sustainable development. CSA provides the possibility to help countries achieve sustainable development and national contribution goals. The combination of the implementation of the "Transforming Our World: The 2030 Agenda for Sustainable Development" and CSA provides an opportunity to improve the overall sustainability of CSA achievements and promotes the integration of CSA with other sustainable development strategies. The interconnection of goals between CSA and sustainable development needs to be fully understood, to better achieve the sustainable development goals contained in "The 2030 Agenda for Sustainable Development" (FAO, 2019).

Agriculture is an important part of national decision-making and achieving sustainable development goals: Agriculture is a key driving force for social development, especially in developing and underdeveloped countries, where the output of the agricultural sector often accounts for more than 30% of gross domestic product (GDP) (World Bank, 2019). FAO (2016a) noted that food and agriculture are the core of the "Transforming our World: The 2030 Agenda for Sustainable Development". The core of sustainable development is to eradicate poverty and hunger, maintain the sustainable use of natural resources, and solve climate change problems. In 2015, countries adopted this agenda, focusing on the most urgent sustainable development priorities, including eradicating poverty and hunger, reducing inequality, tackling climate change, creating decent jobs, and economic growth. The Paris Agreement also emphasized the importance of agriculture in its contribution to national decision-making. FAO (2016b) analyzed the prospective donations identified by 22 national data centers and 140 countries, and the results showed that 131 countries prioritized the adaptation and/or mitigation actions of the agricultural sector, and nearly 95% of developing countries mentioned the issue of adaptation in the agricultural sector. Agriculture has a high priority during development in most countries, with 71% of developing countries and 98% of developed countries mentioning agricultural development as a national priority (FAO, 2016b). In addition, the Paris Agreement emphasizes on the importance of action on climate change for achieving "the fundamental priorities of guaranteeing food security and ending hunger" (COP, 2015).

CSA is the key to achieving sustainable development goals and national decision-making contributions: Climate change, agricultural production, natural resources, ecosystems, and food security are important elements of the sustainable development goals, which are consistent well with the goals of CSA (FAO, 2007). CSA will play an important role in meeting these challenges. Due to its multiobjective nature, CSA provides the possibility of achieving more sustainable development goals, rather than focusing on agriculture or climate intervention with a single focus. The adaptation and mitigation of climate change in agricultural production are an important part of the national sustainable development (FAO, 2016b). The frequent occurrence of extreme climatic events such as sea-level rise, floods, and droughts accelerated by the greenhouse effect is the "new normal" facing humankind in the future, which poses a greater challenge to the realization of several goals of sustainable development. Relevant departments and policymaking organizations need to fully consider the potential impacts of climate change (Ziervogel *et al.*, 2008). CSA provides abundant opportunities to cope with the impact of future climate change, especially it has shown great potential for advancing sustainable development goals. CSA emphasizes on sustainably increasing agricultural productivity and farmer incomes, adapting and building the ability of agriculture to cope with climate change, and reducing and/or removing greenhouse gas emissions as possible, which provide a great opportunity to solve the increasingly serious problems of food security and global poverty. Therefore, the development of CSA should be a priority for countries to achieve sustainable development. At the same time, it is necessary to carry out strategic planning for CSA to enhance the synergy of the agricultural sector in formulating development strategies, plans, and projects, and reduce the trade-offs among interrelated goals such as sustainable production, adaptation to climate change, and mitigation of climate change, while taking into account the national development priorities and the essential needs and actual conditions of regions (FAO, 2005).

2.3 Development Status of International CSA

2.3.1 *Main Content of International CSA*

The research and practice of CSA across the world cover a wide range and basically accord with the concept of CSA, focusing on the goals of improving adaptability, reducing greenhouse gas emissions, sustainably increasing productivity, and bringing in benefits in the context of climate change (Lipper *et al.*, 2009). It is carried out in related fields of planting and breeding production, land management, resource and energy utilization, and ecological governance. International CSA mainly includes the construction of three important systems (FAO, 2008a, 2008b).

The first is the construction of climate-smart crop production systems, focusing on the goals of high yield, intensification, flexibility, sustainability, and low emission, and exploring feasible ways to improve the overall efficiency, resilience, adaptability, and emission reduction potential of the production system through crop production system optimization and technical improvement. Based on the analysis of the constraints and reasons for developing CSA in the current production system, a feasible optimization approach and corresponding technical support are proposed, and the main technologies include cultivating new varieties that are adaptable to climate change, improving water and fertilizer resource utilization efficiency, developing sustainable soil management technologies, developing new integrated pest control technologies, improving agricultural information prediction technologies, and developing greenhouse gas emission reduction technologies.

The second part is the system and policy improvement. In national-level policies and action plans, it is necessary to comprehensively consider agricultural development, adaptation, and mitigation of climate change and then formulate a national action plan and related promotion policies to address climate change. In the joint actions of departmental coordination and relevant stakeholders, resource management is integrated around the objectives and tasks, and departments are coordinated and advanced to encourage all parties to participate together. In addition, the system reform and innovation mechanism are mainly realized from the following aspects: formulating relevant regulations and standards from production to sales, promoting rapid information dissemination and resource sharing, improving service mechanisms of agricultural technology extension, building farmers' participation capacity, and reforming finance and insurance systems.

The third part is finance and capacity building. CSA harmonizes food security, while agricultural development and climate change is not cost-free, which requires large-scale investment to meet its needs. The formulation, research, and extension of CSA policies at the national level, including related capacity building and support for farmers' actions, require sufficient investments. Therefore, the development of CSA faces challenges in financing channels and innovation mechanism.

In January 2012, the first batch of projects of CSA, jointly implemented by FAO and the European Union (EU), were launched in Malawi, Vietnam, and Zambia, raising €5.3 million for the purpose of promoting climate-smart farming practices and carrying out research on supporting issues such as policy support. In recent years, the Global Environment Facility (GEF) has supported developing countries such as Ecuador, Guinea, Mali, Senegal, and Cambodia to carry out CSA projects for dealing with agricultural risks of climate change and soil degradation, developing conservation agriculture such as less tillage, permanent soil cover and rotation, and building climate-smart value chains. In 2014, the launch of the Italian Ministry for the Environment, Land and Sea (IMELS) CSA project and the establishment of the Global Alliance for CSA pushed the research and application of global CSA to a new peak.

2.3.2 *International CSA Actions*

The GEF has set up key areas of climate change and allocated special funds around the goals of the United Nations Framework Convention on Climate Change to support developing countries to carry out research on the mitigation and adaptation to climate change and reduce greenhouse gas emissions, create benefits by improving local economy and environmental conditions, and realize the transformation of energy market in developing countries (FAO, 2015).

CSA is also one of its main funding directions mainly manifested in three aspects: (1) achieving sustainable land management through innovative financial and market mechanisms, ensuring the carbon sequestration and emission reduction of small-scale farmland; (2) improving the management of agriculture and grassland by maintaining soil organic matter; and (3) increasing the availability of functional technologies for ecological services of crops, trees, and animal husbandry, and expand the scope of CSA (Guan *et al.*, 2017a).

The World Bank Group (WB) is committed to cultivating CSA. In the World Bank's Climate Change Action Plan, it promised to establish an early warning system for 100 million people, assist at least 40 countries in making CSA investment plans, continue to develop measurements of related project output and indicators as mainstream, and incorporate GHG emissions into related projects. The World Bank's Global Practice Development Bureau has developed a global CSA indicator system (McCarthy, 2014), including policy indicators, technical indicators, and outcome indicators. It can help evaluate the development situations of CSA in different countries by collecting and comprehensively evaluating global agricultural production data. And it is helpful to the wider promotion of CSA in the world.

In 2014, FAO and IMELS launched a project, signed a memorandum of understanding to strengthen international coordination and cooperation on CSA, and actively established the Global Environment Bilateral Trust Fund among different stakeholders including countries, international organizations, nongovernmental organizations, and farmers. Based on the scientific development of CSA, the project supports specific activities to determine the knowledge needs and the actions to be taken in CSA. The project aims to ensure the farmers' livelihood and food security by developing natural resource management and utilization measures, and adopting appropriate methods and technologies to produce, process, and sell agricultural products. Farmers and their knowledge should be regarded as important custodians of natural resource heritage, so they can improve the adaptability of the landscape while improving the resilience of ecosystems and communities. The main output of the project includes supporting information and knowledge sharing through multiple communication channels, capacity building, and expansion, and through the production and dissemination of FAO knowledge products for South–South cooperation. Stakeholder consultation, shared lessons learned, and information sharing are other core elements of project enhancement. Based on the introduction of pre-feasibility study cases of active bilateral cooperation in Botswana, Ecuador, and Ethiopia, it is hoped that further development in the field of climate change will be achieved to explore the possibility of establishment of CSA projects in developing countries and strengthen national capacity building and agricultural expertise. The project focuses on local agricultural expertise and traditional knowledge to develop strategies to adapt and mitigate climate change to reduce the impact of climate change, while providing policy support, technical assistance, and financial support. Assessing local potential and limitations is a solution to better solve the problem of adaptation and mitigation of climate change. The beneficiaries include not only policymakers, technical consultants, and researchers but also other rural communities and producers who are vulnerable to climate change. The successful implementation of the IMELS-FAO CSA Project is also the core force that promotes the establishment of the Global Alliance for CSA.

2.3.3 *International CSA Organizations*

The Global Alliance for Climate-Smart Agriculture (GACSA) led by the United States was established in New York on September 24, 2014. The first members include 18 countries (United States, Canada, United Kingdom, France, the Netherlands, Spain, Mexico, Nigeria, Japan, Philippines, Vietnam, etc.) and 32 international research institutions [FAO, the Consultative Group on International Agriculture Research (CGIAR), the International Fertilizer Industry Association (IFA), etc.].

The GACSA has formulated corresponding development goals and plans that include sustainable and equitable increases in agricultural productivity and farmer incomes, increasing resilience of food systems and agricultural livelihoods, and reducing agricultural greenhouse gas emissions. The GACSA is an inclusive, voluntary, and action-oriented CSA platform, whose goal is to ensure food security, enrich food nutrition, and enhance the ability to respond to climate change. The GACSA is committed to achieving the three goals of CSA: (1) increase agricultural productivity and farmer incomes by a sustainable manner; (2) enhance farmers' ability to cope with extreme weather and climate change; and (3) minimize greenhouse gas emissions related to agriculture as possible. Specific priorities and solutions need to be consistent with national policies and priorities. The GACSA aims to promote and help to establish transformational partnerships to encourage actions and methods that reflect CSA's three goals: productivity, adaptation, and mitigation. The duty of GACSA is to address the challenges to food security and agricultural production by leveraging the wealth of resources, information, and expertise among its members to promote concrete actions at all levels. It provides a communication platform for people engaged in CSA to discuss, exchange experiences, information, and opinions. Specifically, agricultural sectors need to adopt measures related to climate change and greenhouse gas emissions.

The Africa Climate-smart Agriculture Alliance (Africa CSA) was established at the UN Climate Change Conference in June 2014. Its goal is to support 6 million farmers in sub-Saharan Africa to adopt climate-smart practices and methods by 2021 to improve the lives and livelihoods of small farmers and to ensure the fairness and sustainability of agricultural system development (Swallow *et al.*, 1999). At the United Nations Climate Summit in September 2014, the United States proposed the establishment of the North America Climate-Smart Agriculture Alliance (NACSAA), the main purpose of which is to provide farmers and their value chain partners with a platform to jointly develop methods to improve the flexibility of agricultural production systems, to slow down and adapt to climate change, and to increase agricultural flexibility under the risk of climate change in the future.

2.3.4 *International CSA Development Experience*

After more than 10 years of development, countries around the world have recorded significant achievements and accumulated abundant experiences that can provide lessons for the future development of CSA. Accumulated experience could be summarized as follows:

Firstly, countries have formulated detailed climate-smart agricultural development goals and implementation plans. Developed countries such as the United States and the European Union, or some developing countries such as Brazil and Ethiopia, have proposed practical goals in line with their national development level and have released detailed support guidelines for core areas to help realize the transition to CSA.

Secondly, countries focused on the integration and application of new materials, technologies, and methods in climate-smart agricultural practices. Current agricultural research achievements in new materials for carbon sequestration and emission reduction (e.g. nitrification inhibitors and emission reduction feed additives), new technologies (e.g. farming technology, agricultural models, and remote sensing technology), and new methods (e.g. life cycle analysis methods and information management systems) have provide necessary support for the implementation of CSA.

Thirdly, projects are supported by long-term and stable funding. CSA is a new type of agricultural development model, and most countries have launched a number of projects and provided stable funds for long-term support on research and application of CSA.

Fourthly, a globalized method of cooperation was formed. It has been realized that the sustainable development of agricultural production requires the joint efforts of all countries in the world. Moreover, FAO, WB, the United States, the European Union, and other developed countries, regions, and organizations have advantages in terms of technology and funding. In recent years, they have provided specific support and assistance for CSA research in developing countries through various forms of international cooperation (Guan *et al.*, 2017b).

References

Adenle A A, Azadi H, Arbiol J. 2015. Global assessment of technological innovation for climate change adaptation and mitigation in developing world. *Environmental Management*, 161: 261–275.

Barrett C, Barrett B, Carter B, *et al.* 2007. Poverty traps and climate and weather risk: Limitations and opportunities of index-based risk financing. IRI Technical Report.

Burke M, Emerick K. 2016. Adaptation to climate change: Evidence from US agriculture. *American Economic Journal: Economic Policy*, 8 (3): 106–140.

Celeridad R L. 2018. *Contextual and Universal: Scaling Context-Specific Climate Smart Agriculture.* World Agroforestry Centre, Nairobi, Kenya.

Deryng D, Conway D, Ramankutty N, *et al.* 2014. Global crop yield response to extreme heat stress under multiple climate change futures. *Environmental Research Letters*, 9 (3): 034011.

FAO. 2007. *The State of Food and Agriculture, Paying Farmers for Environmental Services.* Food and Agriculture Organization, Rome, Italy.

FAO. 2008a. Institutions to support agricultural development. Unpublished Report, FAO, Rome, Italy.

FAO. 2008b. The Investment Imperative, paper from the FAO High Level Conference on World Food Security: The Challenges of Climate Change and Bioenergy. Food and Agriculture Organization of the United Nations, Rome, Italy. https://www.fao.org/fileadmin/user_upload/foodclimate/statements/fao_diouf_e.pdf.

FAO. 2009a. *Food Security and Agricultural Mitigation in Developing Countries: Option for Capturing Synergies.* Rome, Italy.

FAO. 2009b. *The State of Food and Agriculture — Livestock in the Balance.* Rome.

FAO. 2009c. *Increasing Crop Production Sustainably, the Perspective of Biological Processes.* Food and Agriculture Organization of the United Nations, Rome, Italy.

FAO. 2009c. Increasing Crop Production Sustainably, the Perspective of Biological Processes. Rome. Food and Agriculture Organization of the United Nations, Rome, Italy. https://www.fsnnetwork.org/sites/default/files/increasing_crop_production_sustainably--the_perspectives_of_biological_process.pdf.

FAO. 2009d. *The Investment Imperative, paper from the FAO High Level Conference on World Food Security: The Challenges of Climate Change and Bioenergy.* Food and Agriculture Organization of the United Nations, Rome, Italy.

FAO. 2015. *The Economic Lives of Smallholder Farmers: An Analysis Based Onhousehold Data from Nine Countries.* Food and Agriculture Organization of the United Nations, Rome, Italy.

FAO. 2016a. *Food and Agriculture: Key to Achieving the 2030 Agenda for Sustainable Development.* Food and Agriculture Organization of the United Nations, Rome, Italy.

FAO. 2016b. The Agriculture Sectors in the Intended Nationally Determined Contributions: Analysis. Environment and Natural Resources Management Working Paper 62. Rome, Italy.

FAO. 2018a. *Transforming Food and Agriculture to Achieve SDGs.* Food and Agriculture Organization of the United Nations, Rome, Italy.

FAO. 2018b. *World Livestock: Transforming the Livestock Sector Through the Sustainable Development Goals.* Rome, Italy. Licence: CC BY-NC-SA 3. 0 IGO. Food and Agriculture Organization of the United Nations, Rome, Italy.

FAO. 2019. *Climate-Smart Agriculture and the Sustainable Development Goals: Mapping Interlinkages, Synergies and Trade-offs and Guidelines for Integrated Implementation.* Food and Agriculture Organization of the United Nations, Rome, Italy.

Guan Dahai, Zhang Jun, Zheng Chengyan, *et al.* 2017a. Overview and reference of foreign climate smart agriculture development. *World Agriculture*, 4: 23–28.

Guan Dahai, Zhang Jun, Wang Qingmei, *et al.* 2017b. Climate-smart agriculture and its enlightenment to my country's agricultural development. *China Agricultural Science and Technology Review*, 19 (10): 7–13.

Hansen B, Thorling L. 2008. Review of survey activities 2007: use of geochemistry in groundwater vulnerability mapping in Denmark. *Geological Survey of Denmark, Greenland Bulletin*, 15 (15): 45–48.

Howden S M, Soussana J F, Tubiello F N, *et al.* 2007. Adapting agriculture to climate change. *Proceedings of the National Academy of Sciences of the United States of America*, 104 (50): 19691–19696.

IPCC. 2001. *An Assessment of the IPCC. Synthesis Report -Summary for Policymakers.* Cambridge, UK and New York, USA: Cambridge University Press.

IPCC. 2007b. Climate change 2007: Mitigation. In B Metz, O R Davidson, P R Bosch, *et al.* (eds.) Contribution of Working Group III to the Fourth Assessment Report of the IPCC. Cambridge, UK and New York, USA, Cambridge University Press.

IPCC. 2013. Climate Change 2013: The physical science basis. Working Group I contribution to the Fifth Assessment Report of the Intergovernmental Panel on Climate Change, Cambridge University Press, Cambridge, UK.

IPCC. 2014. Climate change 2014: Mitigation of climate change. In O Edenhofer, R Pichs-Madruga, Y Sokona, *et al.* (eds.) Contribution of Working Group III to the Fifth Assessment Report of the Intergovernmental Panel on Climate Change. Cambridge, UK, Cambridge University Press, and New York, NY, USA.

Kahiluoto H, Kaseva J, Balek J, *et al.* 2019. Decline in climate resilience of European wheat. *Proceedings of the National Academy of Sciences*, 116 (1): 123–128.

Lipper L, McCarthy N, Zilberman D, *et al.* 2017. Climate smart agriculture-building resilience to climate change. *Natural Resource Management and Policy.*

Lipper L, Sakuyama T, Stringer R, *et al.* (eds.). 2009. *Payment for Environmental Services in Agricultural Landscapes: Economic Policies and Poverty Reduction in Developing Countries, Natural Resource Management and Policy Series*, Vol. 31. Springer. London, UK.

Lobell D B, Hammer G L, McLean G, *et al.* 2013. The critical role of extreme heat for maize production in the United States. *Nature Climate Change*, 3 (5): 497–501.

Lobell D Schlenker W, Costa-Roberts J. 2011. Climate trends and global crop production since 1980. *Science*, 333 (6042): 616–620.

Swallow B, McCarthy N, Swallow B, *et al.* 1999. Property rights, risk, and live-stock development in Africa: Issues and project approach. *International Symposium on Property Rights, Risk.*

Mann W, Lipper L, Tennigkeit T, *et al.* 2009a. Food Security and Agricultural Mitigation in Developing Countries: Option for Capturing Synergies. Food and Agriculture Organization of the United Nations, Rome, Italy. https://www.fao.org/3/i1318e/i1318e00.pdf

McCarthy N. 2014. *Climate-smart Agriculture in Latin America: Drawing on Research to Incorporate Technologies to Adapt to Climate Change.* Inter-American Development Bank, Washington, USA.

Moore F C, Lobell D B. 2014. Adaptation potential of European agriculture in response to climate change. *Nature Climate Change*, 4 (7): 610–614.

Olesen J E, Trnka M, Kersebaum K C, *et al.* 2011. Impacts and adaptation of European crop production systems to climate change. *European Journal of Agronomy*, 34 (2): 96–112.

Raney T, Steinfeld H, Skoet J. 2009b. The State of Food and Agriculture -- Livestock in the Balance. Food and Agriculture Organization of the United Nations, Rome, Italy. https://www.fao.org/policy-support/tools-and-publications/resources-details/en/c/1235525/.

Ray D K, Gerber J S, MacDonald G K *et al.* 2015. Climate variation explains a third of global crop yield variability. *Nature Communications*, 6: 5989.

Roy J, Tschakert P, Waisman H, *et al.* 2018. Sustainable Development, Poverty Eradication and Reducing Inequalities. In Global Warming of 1.5°C: An IPCC Special Report on the impacts of global warming of 1.5°C above pre-industrial levels and related global greenhouse gas emission pathways, in the context of strengthening the global response to the threat of climate change, sustainable development, and efforts to eradicate poverty IPCC. https://www.ipcc.ch/sr15/chapter/chapter-5/".

Schlenker W, Roberts M. 2009. Nonlinear temperature effects indicate severe damages to U. S. crop yields under climate change. *Proceedings of the National Academy of Sciences of the United States of America*, 106 (37): 15594–15598.

SEI. 2008. *Climate Change and Adaptation in African Agriculture.* Stockholm Environment Institute.

Smit B, Wandel J. 2006. Adaptation, adaptive capacity and vulnerability. *Global Environmental Change*, 16 (3): 282–292.

Trnka M, Rotter R P, Ruiz-Ramos M, *et al.* 2014. Adverse weather conditions for European wheat production will become more frequent with climate change. *Nature Climate Change*, 4 (7): 637–643.

United Nations Environment Programme (UNEP). 2010. Assessing the environmental impacts of consumption and production: priority products and materials. In E Hertwich, E van der Voet, S Suh, *et al.* (eds.) A Report of the Working Group on the Environmental Impacts of Products and Materials to the International Panel for Sustainable Resource Management.

Webber H, Ewert F, Olesen J E, *et al.* 2018. Diverging importance of drought stress for maize and winter wheat in Europe. *Nature Communications*, 9 (1): 1–21.

World Bank (WB). 2019. Agriculture, forestry, and fishing, value added (% of GDP). https://data.worldbank.org/indicator/NV.AGR.TOTL.ZS?name_desc=true.

Ziervogel G, Cartwright A, Tas A, et al. 2008. Climate Change and Adaptation in African Agriculture. Stockholm Environment Institute. https://www.fao.org/fileadmin/templates/agphome/scpi/SHARP/Climate_change_and_adaptation_in_African_agriculture.pdf.

Chapter 3

International Climate-Smart Agriculture Practice Approach

Abstract

Climate change has led to significant negative impacts on agriculture and food security in many regions, especially in developing countries that highly rely on dryland farming. The Fifth Assessment Report from IPCC found that tropical regions are most sensitive to climate change, where most of the poor and agriculture-dependent populations reside. Increasing the growth rate of agriculture in these regions is essential for achieving poverty eradication and the goal of meeting growing food demand. The idea of climate-smart agriculture (CSA) has gained considerable attention around the world. The goal of CSA is to incorporate adaptation to and mitigation of climate change particularities into sustainable agricultural development policies, programs, and investment strategies. The principles and practices of CSA should follow the requirements of sustainable agriculture and food systems. FAO published a set of guidelines and guidance for the sustainable development of agriculture and food production systems, including: (1) improving the efficiency of resource utilization; (2) conserving, protecting, and restoring nature resources; (3) protecting and improving rural livelihoods; (4) enhancing the resilience of people, ecosystems, and communities; and (5) building effective governance mechanisms. The emphasis of CSA is on evaluating the trade-offs and synergies among its three main goals and remove the obstacles to CSA. CSA actually solves one of the most important issues in sustainable agriculture: what are the needs of agricultural production practices to achieve large-scale transformation? The trade-off between multidimensional goals of CSA is an improvement to many sustainable agricultural methods. However, confusion persists in the concept and theoretical basis of CSA, and practical experience is relatively lacking at the national level. This chapter systematically organizes the main practical approaches of CSA and provides support for further improvement of CSA theory and technical system. The content includes the innovations in technology, management, and systems of CSA, methods of climate vulnerability assessment, and their effects, and the methods and approaches to reduce agricultural greenhouse gas emissions and the development strategy of CSA.

3.1 Innovations of CSA

Innovation in agricultural management strategies is an important countermeasure to effectively adapt to and mitigate climate change. The impacts of climate change on agriculture production vary with time and space with high uncertainties. The implementation of CSA requires great resilience to climate change in the agricultural production system, as well as high resource utilization efficiency in adapting to and mitigating climate change. We need to rethink how to promote innovation to deal with the heterogeneous and uncertain impacts of climate change. In the course of achieving CSA in developing and developed countries, innovation will be the key. Innovative technologies play a key role in crop variety adaptation, improving both the efficiency of water and fertilizer, and the sustainability of agricultural production, while innovations on management methods and mechanisms are conducive to enriching adaptive methods and forms to climate change, and effectively increase the adoption rate of high-tech. Besides, it can help prevent or promote the migration of production and people, increase trade and aid, and improve insurance efficiency and inventory feasibility, etc., and it is more important in dealing with the heterogeneity and uncertainty of climate change.

3.1.1 *Technological Innovations*

(1) Cultivating new cultivars that adapt to climate change well: Due to the rise in temperatures and increased climate variability, it will be very important to develop new crop cultivars that can withstand these changes.

As the frequency of climate change increases, it will also be important to detect changes and develop genetic material that can quickly adapt to such changes.

(2) Technology to improve the utilization efficiency of water and fertilizer: There is often a big gap between theoretical and actual resource utilization of different agricultural technologies in the field. For example, water utilization efficiency is only around 50% during flood irrigation, which can be increased to 90% by drip irrigation. Previous studies have shown that the adoption of efficient resource utilization technologies tends to increase production, save inputs, and reduce pollution (Khanna and Zilberman, 1997). Technological innovation can reduce the use of water, fertilizers, and chemicals, improve resource utilization efficiency, and reduce greenhouse gas emissions.

(3) Developing sustainable soil management technologies: Agricultural practices in developing countries often lead to a decline in soil quality, and extreme weather may make this problem worse. Sustainable soil management aims to increase yields without reducing the quality and quantity of soil and water resources, and at the same time increase the carbon sequestration capacity of the soil. Improving existing practices to achieve sustainable soil management has great significance. The existing sustainable soil management technologies in production include applying more organic fertilizers, reducing soil disturbance, and constructing agroforestry systems.

(4) Developing new integrated pest control technologies: The migration of pests urges people to develop new pest treatment technologies that can achieve comprehensive effects, such as environmental friendliness, high cost-effectiveness, convenient use, and significant efficiency. The development of this technology requires the use of multidisciplinary methods such as biological, mechanical, chemical controls, and genetics. Currently, researchers are working on identifying new and emerging pest problems, which will facilitate the development of innovative pest control methods.

(5) Improving agricultural information forecasting technology: Weather forecasting has a profound impact on agricultural production, and it directly affects farmers on the application of irrigation and pest control technologies. Besides, it helps to increase yield and resource utilization efficiency (Parker and Zilberman, 1996). Climate change will intensify in the future, and agricultural production will face greater weather problems. Certainty and reliable weather forecasts are especially important. The establishment of necessary weather stations and transmission systems is an important way to upgrade the agricultural weather forecasting technology.

(6) Developing greenhouse gas emission reduction technologies: Reducing greenhouse gas emissions is the key to effectively mitigate climate change, and it is also an important goal of CSA. Greenhouse gas reduction technologies include innovative cultivation management measures, improved crop varieties, using biochar, and protective agricultural technologies. Using high-yield varieties and protective technologies to reduce the negative environmental effects of the land, atmosphere, and fossil fuels of the agricultural ecosystem is an important greenhouse gas mitigation strategy (Lal, 2011).

(7) Improving farm storage facilities and extending the shelf-life of agricultural products: Parfitt *et al.* (2010) found that, in developing countries that lack basic storage capacity, the postharvest losses of farms are very large, and most of the losses happen to self-sufficient farmers. Innovative farm storage infrastructure helps to solve the production loss caused by increased temperature and the frequency of earthquakes. In the context of the heterogeneity of climate change, designing a reliable system that is reasonably priced and easy to install and operate is the biggest challenge. Increasing yield per unit area often reduces the agricultural production footprint, which is conducive to reducing production losses caused by climate change. Increasing the shelf-life in the context of climate change can help reduce the risk of deterioration caused by temperature rise. A longer shelf-life will reduce transportation and storage costs, especially for wastes related to agricultural industrial distribution.

3.1.2 *Management Innovations*

(1) Optimizing the time and amount of input, and significantly improving the efficiency of the agricultural water or chemical inputs by adopting information-intensive management practices: Dobermann *et al.* (2004) recommended using precise management techniques to save investment and increase yield. The integrated technology of water and fertilizer is the key to improving the utilization efficiency of water and fertilizer, and it has been widely adopted in developed countries such as the United States. However, the utilization efficiency of water and fertilizer resources in developing countries is still low. It is a special challenge to develop precise management technologies for resource-poor developing countries.

(2) Integrated pest management (IPM) will improve the effectiveness of pest management in detecting and coordinating pest control activities and help effectively deal with the increased pest pressure caused by climate change: The effectiveness of combating climate change will benefit from the development of affordable and easy-to-implement comprehensive prevention strategies. IPM emphasizes the measurement of pest stress and combines with other methods (chemical, genetic modification, and biological) to optimize the net benefits of treatment, and at the same time considers the dynamics of pests and environmental side effects. The application of IPM is limited by the cost of pest monitoring and the difficulty of tailoring IPM to biological climatic conditions (Waterfield and Zilberman, 2012).

(3) Strengthening land use and farm management: With changes in technology and economic conditions, climate change has exacerbated the challenges of farm management. Both land use and farm management need improved management tools to promote the selection of crop types and varieties, the configuration of crop layouts, production practice choices, and application of technology. The increase in data quality, computing power, and communication capabilities will provide opportunities for the introduction of new management tools.

(4) Improving input use and management and establishing a shared knowledge base for farmers: For example, comprehensive management and improvement of water use efficiency depend on the meteorological information collected and provided by regional weather stations. Formulating strategies to deal with crop diseases and measures to control pest resistance require collective action. Effective land use management should consider the externalities between crops and other production activities in a region.

Therefore, regional cooperation institutions among small producers should be established and developed to provide public products and expand production scale within them, which will bring more benefits for regional agricultural producers. Poteete *et al.* (2010) proposed that various forms of management and decision-making institutions should be built and developed to respond to the needs of various collective actions in the context of climate change. But different situations may require different solutions, and innovative institutional designs should adapt to emerging climate changes.

(5) Enhancing the promotion of agricultural insurance: agricultural insurance is an important starting point for improving risk management strategies under climate change, reducing other preventive activities, thereby reducing the risk of yield loss. Implementation of other adaptation strategies reduces the need for insurance. The implementation of agricultural insurance may require monitoring to overcome the difficulties of adverse selection, and the devising an insurance strategy to deal with climate change must be done with caution (Tol, 2009). The implementation of agricultural insurance may require monitoring to overcome the difficulties of adverse selection, and the devising an insurance strategy to deal with climate change must be done with caution.

(6) Optimizing the flexible management design of the supply chain: The flexible management design of the supply chain is essential to improve its efficiency (Luet *et al.*, 2015). Agriculture in developing countries is undergoing a systemic revolution in food production (Lu *et al.*, 2015). They have introduced reasonable supply chains to better store agricultural products and increase the diversity of products, while at the same time directly linking farmers and supermarkets (Reardon and Timmer, 2012). These kinds of modern supply chains have adopted many innovative practices and have a great potential in dealing with the effects of climate change.

3.1.3 *Institutional Innovations*

The institutional innovation of CSA refers to inventing new ways to deal with matters with a certain background. It occurs at a large scale and in agricultural management systems, including systems that can initiate the innovation process and that allow the implementation of adaptation strategies. Adaptation to climate change, including the mitigation of climate change, requires a high degree of innovation in physical technology as well as institutions and policies. A key factor in formulating climate change adaptation policies is to strengthen scientific research and international cooperation investment and build awareness of the benefits of climate change and adaptation. To implement system innovation, it is necessary to expand the scale of production and operation, and then test at the implementation level. A sound marketing and education plan is required to provide innovative ideas for practitioners. The implementation of CSA requires innovative network design to extend technology from scientists to producers and operators. This expansion should include not only public extension services, but also private companies, cooperatives, and nongovernmental organizations. Good management and flexible mechanisms can strengthen the cooperation of leading institutions with various adaptation strategies and reflect the corresponding value in the agricultural sector's response to climate change (FAO, 2016).

3.2 Accurate Climate Vulnerability Assessment is an Important Prerequisite for Formulating Strategies to Deal with Climate Change

3.2.1 *Crop Production Monitoring Based on Satellite Remote Sensing Technology*

Satellite-based remote sensing technology is an important means to monitor the dynamics of water resources and the status of agricultural production to assess regional food security. In recent years, satellite observation has been widely used to mitigate the adverse effects of extreme events and promote resistance and recovery capabilities to the effects of climate change. Population growth and increasing resource demand in developing countries will increase the vulnerability of agricultural production, which may be exacerbated by climate change. Objective satellite data can be used as (1) detectors of crop growth conditions and food supply; (2) a risk management index to reduce the impact of crop failure; (3) important educational and information tools in crop selection, resource management, and other adaptation or mitigation strategies; (4) an important tool on food aid and transportation; and (5) a water resource allocation management tool. Satellite data has huge value in monitoring crop production, food security, river flow, and river basin planning in many areas. These data can be used as valuable intelligent decision-making tools for CSA. The advantages of satellite observation are as follows: (1) It provides disaster warnings for the available grain output, thereby reducing the risk of production reduction; (2) Humidity and temperature anomalies can be used as indicators for insurance projects to compensate farmers' losses; (3) The reports of growth conditions can be used to identify the return period of different levels of crop failure and can also be used to differentiate different levels of crop failure harvest vulnerabilities, which are important information for insurance industry to calculate insurance premiums; (4) It uses climatology to determine the feasibility of alternative crop production, which may surpass the original crops grown in the region. Multi-crop system is an effective measure to prevent catastrophic crop failures, and it may be an important supplement for climate change mitigation activities, agricultural productivity, climate resilience, and natural resource management (Larson *et al.*, 2015). When monitoring crop yields, yield prediction models need to be specifically calibrated for each crop and specific location. Specifically, the production forecast model is calibrated based on historical values using linear changes in temperature and humidity abnormal values as predictive factors. When monitoring river flow, quantitative and independent measurement of river flow is essential for water conservancy and applying right sand allocation plans (Dinar *et al.*, 2010), which help to rationally plan and allocate river basin water to each water sector. In addition, satellite remote sensing technology is also widely used in crop phenotype monitoring to provide a technical platform support for crop breeding to cope with climate change.

3.2.2 *Comprehensive Evaluation Method of Crop Model*

The evaluation of the prediction effect of climate change requires a multidimensional method to evaluate its performance in terms of economic, environmental, and social benefits in different agricultural systems. CSA combines the three goals — climate change adaptation, mitigation, and restoration — into the framework of a sustainable agricultural system. At present, there is no clear and systematic method to achieve the comprehensive goals of CSA. The Agricultural Model Intercomparison and Improvement Project (AgMIP) developed a Regional Integrated Assessment (RIA) method to comprehensively assess the impact of climate change and ways to adapt to and mitigate climate change, which provides a framework for CSA assessment (AgMIP, 2015; Antle *et al.*, 2015). The method evaluates and quantifies the performance of the agricultural production system indicators determined by stakeholders and scientists and evaluates the vulnerability and resilience of the production system to climate change, so as to improve the ability of the agricultural system. By simulating the impacts of future climate scenarios and socio-economics on agricultural production, how a system responds to climate and other changes is assessed, including system changes for climate adaptation and mitigation. This predicts the contribution of production technology advancements to agricultural production and optimizes adaptation measures. AgMIP regional integrated evaluation method is based on stakeholder discussion and research evaluation literature and comprehensively considers agricultural production, population poverty, food security, environment, and other indicator standards, and uses model simulation ideas to evaluate climate change impacts and effect of system adaptation. AgMIP regional integrated evaluation method aims to enable the research team to cooperate with stakeholders to solve four core questions: (1) How sensitive is the current agricultural production system to climate change? (2) What is the impact of adaptation strategies in the current world? (3) How climate change affects the future agricultural production system? (4) What are the benefits of adapting to climate change? AgMIP regional integrated evaluation methods are widely used to support technological development. For example, it has been used to promote farm-specific agricultural interventions, design and evaluate safety nets tailored to local conditions, food security, or market-oriented interventions.

3.2.3 *Climate Vulnerability Assessment Based on Mathematical Statistical Models*

Mathematical statistical models based on long-term historical statistical data, such as linear models and maximum entropy models, are widely used in climate change vulnerability assessment and is an important tool for predicting regional climate change on crop production. Climate change is an indisputable fact. Climate change, characterized by climate warming, has led to a significant increase in the frequency of extreme weather events such as drought and high temperature, which significantly affect the production of major grain crops such as rice, wheat, and corn (Schlenker and Roberts, 2009). Through historical statistical data mining, there is great significance to predict the impact of climate change on future crop production. Previous studies have shown that high-temperature events significantly affected crop yields in the past few decades. Though corn production in the Midwestern United States has increased at a rate of 17% per decade due to the steady progress in technology, climate change will produce an increasingly greater back-phagocytic effect.

Nonlinear model analysis showed that corn yield will decrease by 4.2%, 21.8%, and 46.1% under the uniform temperature rise scenarios of 1°C, 3°C, and 5°C, respectively (Lobell *et al.*, 2011). Only increasing corn yield at 6.6% every 10 years can offset the adverse effects of temperature rise by 3°C in the next three decades. China's domestic research showed that an increase in daily minimum temperature is beneficial to increasing corn yield (Tao *et al.*, 2006). The raised temperature increases the accumulated temperature in the northern part of the Northeast farming area, reduces the risk of low temperature and cold damage, and increases corn yield (Meng *et al.*, 2013). Studies have shown that the daily minimum temperature in May and September in the northeastern farming area is significantly correlated with the yield of corn. The increase of daily minimum temperature by 1°C will increase the yield of corn by 284–303 kg hm^{-2} (Chen *et al.*, 2011).

3.3 Improving Adaptability of Agricultural Systems to Climate Change

3.3.1 *Definition and Assessment of Adaptation*

The impact of climate change on crop production systems is multifaceted. Adaptation is an option to reduce the adverse effects of climate change on crop production systems. Appropriate adaptation technologies can significantly reduce the adverse effects of climate on crops and provide new opportunities for crop production (Smit and Wandel, 2006; Howden *et al.*, 2007; Thornton *et al.*, 2010; Reidsma *et al.*, 2010; Tao and Zhang, 2010; Olmstead and Rhode, 2011). Adaptation to climate change refers to the adjustment of natural or artificial systems to enhance their ability to adapt to anticipated stimuli or effects of climate change (IPCC, 2001). According to the different perspectives of adaptive measures, adaptation can be divided into many types, including initiative adaptation and automatic adaptation, which are also respectively called long-term adaptation and short-term adaptation (IPCC, 2001; Smit and Wandel, 2006; Olesen *et al.*, 2011). Initiative adaptation refers to the government's adjustment of relevant policies to make climate change adaptation measures more effective. The government's adaptation measures to address climate change mainly include strengthening infrastructure construction, increasing investment in resistance breeding, improving disaster resistance systems and regulations, providing disaster risk subsidies, etc. (Howden *et al.*, 2007). Automatic adaptation means that farmers adopt technical adjustments in response to climate change (Olesen *et al.*, 2011). Farmers adopt several measures to adapt to the impact of climate change on crop production incorporating adjusting crop-sowing dates, selecting more suitable crop varieties, changing soil cultivation techniques, optimizing water and fertilizer management, and strengthening plant diseases and pest control (Smit and Skinner, 2002; Challinor *et al.*, 2007; Tubiello *et al.*, 2007; Wang *et al.*, 2010a, 2010b; Kato *et al.*, 2011; Olesen *et al.*, 2011; Olmstead and Rhode, 2011). The actual impact and vulnerability of climate change are closely related to the characteristics of farmers, which will affect the use of management measures and adaptation technologies (Reidsma *et al.*, 2007, 2009, 2010; Bryan *et al.*, 2009; Thorton *et al.*, 2010). Reidsma *et al.* (2010) stated that farmers owning different scales of farms in Europe respond differently to climate change. Deressa *et al.* (2009) showed that the adaptation technologies that farmers choose to respond to climate change in the Nile River Basin in Africa are significantly related to the specific ecological types of the region.

The factors that motivate farmers to adopt adaptive technologies can be roughly divided into three categories: policy and institutional factors, family characteristics of farmers, and socioeconomic characteristics (Stringer *et al.*, 2009; Huang and Wang, 2014). The results of previous studies indicated that in the event of a disaster, the weather warning, financial support, and technical support given by the government can effectively promote the adoption of various adaptation technologies by farmers (Chen *et al.*, 2014; Tol Richard, 2009). The education level, gender, age, family income of the head of the household, and access to information about climate change will significantly affect the adoption of adaptation technologies by the farmers (Deressa *et al.*, 2009).

At present, there are a few quantitative studies on the anti-disaster effects of adaptation technologies. The research on adaptive capacity is mostly based on crop growth models. The responses of farmers to climate change in the models are mostly hypothetical scenarios, and the simulation effect of the model is somewhat related to actual production gap (Easterling *et al.*, 2000). Therefore, it is of great significance to quantitatively evaluate the effects of adaptation technologies in dealing with extreme weather events.

3.3.2 *Agricultural Insurance and Subsidies Helping Improve Adaptability*

The impact of climate change has led to worsening poverty and food insecurity in rural areas of developing countries (Barrett B B *et al.*, 2007).

Agriculture insurance can effectively mitigate the adverse effects of climate risk on poverty and food security. Insurance is a cost-effective and highly effective means to deal with food insecurity caused by climate change. Good insurance contracts should be priced and executed at a reasonable level, which helps improve farmers' adaptability to climate change. On the one hand, insurance compensation can help families maintain their economic assets and long-term economic viability. To put it more simply, insurance should help families avoid poverty. On the other hand, insurance increases ex post protection, which can produce ex ante effects by increasing the level of expectations and the certainty of return on investment, allowing more families to escape from poverty and food insecurity. Unlike policies that simply transfer food aid to deal with casualties caused by climate shocks, a comprehensive policy that includes insurance factors can address the root causes of food insecurity rather than just symptoms to reduce required total social protection expenditure. The experience from the East African Pastoral Area Index-based Livestock Insurance (IBLI) showed that insurance can generate positive theoretical assessments based on the impact of climate change. Using agricultural insurance to solve climate risks and food insecurity is not an easy solution. Continuous analysis of IBLI experience showed that more work should be done to expand and maintain high-quality insurance contracts. There is still a need for innovation in contract terms and system design, and the establishment of more effective public–private partnerships also requires methodological innovation to regulate agricultural insurance pricing to a certain extent, so insurance can reach its potential and become an era of climate change, which will be a part of the integrated management of social protection and food security.

Future climate change will intensify the volatility of crop yields, and agricultural insurance can play an important role in this context. In developing countries, the assessment of yield losses for single farmer can be very expensive, so insurance contracts are based on local knowledge of the weather. The volatility of the insurance rate of return depends on factors such as weather, so people can regard this insurance contract as a part of protection against agricultural risks. As insurance is a state-owned contract, the payment received by farmers will depend on the probability of a particular weather phenomenon and the amount of insurance purchased. This type of insurance is different from insurance purchased through the use of technology because insurance reduces downside risks while technology reduces downside and upside risks by compressing volatility.

The protection to production project (PtoP) implemented by FAO discussed the role of cash transfer program risk management tools in improving the resilience of sub-Saharan Africa. The results showed that these programs have an important impact on the family's ability to recover. Although the impact on risk management is less uniform, the cash transfer program seems to strengthen community connections, enable households to save and repay debt, and reduce the need to rely on adverse risk response mechanisms. The Zambian case indicated that households receiving cash transfers were much less affected by weather shocks. Among them, the poorest households received the most benefits and food security has improved, although the situation varies from country to country. Therefore, taking the social protection program as a safety net, it can be more effective to consider climate risks in the design and implementation process.

3.4 Climate Change Mitigation

3.4.1 *CO_2 Emission Reduction*

Reducing greenhouse gas emissions from agricultural production systems is one of the core goals of CSA. Protective agricultural technology and water and fertilizer optimization management technology in crop production are important regulatory measures to reduce greenhouse gas emissions from farmland. The Four per Mille Initiative proposed in the Paris Agreement clarified that the method of increasing soil carbon sequestration will reduce greenhouse gas emissions. Soil carbon storage capacity and potential depend on different types of land use and intensity of use in different agroclimatic regions. The rate of decomposition and turnover of soil organic matter mainly depend on the combined effects of soil biota, temperature, humidity, and their chemical and physical compositions. It is also influenced by previous land use and management practices. In the past 50–100 years, farming practices related to soil degradation have reduced the soil organic carbon (SOC) by 13% in many areas (IPCC, 2014). The 13% loss of SOC not only represents a huge loss of water storage (432,000 liters per hectare), but also means that each hectare emits nearly 400 tons of additional carbon dioxide into the atmosphere. The losses of SOC and water-holding capacity are related to practices such as destruction of perennial ground cover, repeated cultivation or continuous grazing, bare fallow farming, removal of crop residues, and burning of grassland. Cash crops and single planting methods that rely on high inputs have achieved the highest possible rate of return and the smallest labor input. However, the price of fossil fuels continues to increase, and applying energy-intensive chemical fertilizers and pesticides is the main source of greenhouse gas emissions. In addition, if used improperly, these chemical inputs can seep into the groundwater, and the resulting water pollution can have serious harmful effects on the ecosystem and human health. Diversified crop rotations and improved fertilizer, seed, and pesticide management technologies can make the use of inputs more effective. Improving efficiency in this area can also potentially reduce greenhouse gas emissions. By improving soil structure and increasing soil biodiversity, no-tillage farming and soil compaction control will also reduce greenhouse gas emissions.

3.4.2 *CH_4 Emission Reduction*

Paddy fields are considered to be the main emission source of CH_4 in agricultural production. Its emissions account for 11% (IPCC, 2013) of the global anthropogenic source CH_4 emissions. It is estimated that the total amount of CH_4 discharged from paddy fields in China reached 6.15 Mt per year, which accounts for about 17.93% of the total domestic annual CH_4 discharge. CH_4 emissions from paddy fields involve three basic processes including CH_4 production, reoxygenation, and transmission to the atmosphere. In an anaerobic environment, methanogenic bacteria act on rice root exudates, sloughs, and soil organic matter and produce CH_4 through anaerobic fermentation. About 30–90% of CH_4 is oxidized by methane-oxidizing bacteria (Bosse and Frenzel, 1997; Ye and Horwath, 2017). The oxidation process of CH_4 in soil is mainly carried out in an aerobic environment. Unoxidized CH_4 enters the atmosphere through bubbles, diffusion, and water aeration tissue of rice plants. Among them, the transmission to the outside world through the aeration tissue of rice is the main method of CH_4 transmission. Water management, straw return, nitrogen fertilizer application, farming measures, use organic fertilizer, and biochar application can affect CH_4 emissions at the paddy field (Zhao *et al.*, 2019).

Water management in the rice-growing season has a decisive influence on CH_4 emissions. CH_4 is produced when methanogenic bacteria act on the methanogenic substrate under strict anaerobic conditions. When the paddy field is flooded, the water layer of the paddy field restricts oxygen transportation to soil from atmosphere, which reduces soil oxidation–reduction potential (E_h). It is conducive to the formation of an anaerobic reduction environment in the soil and provides necessary conditions for the growth and maintenance of methanogens. In the late stage of rice tillering, drainage and baking the field will affect the anaerobic environment required for CH_4 production, and CH_4 will be significantly reduced. Adopting reasonable water management measures (such as intermittent irrigation, controlled irrigation, etc.) can significantly reduce CH_4 emissions in the paddy field.

Water management will also have a great impact on CH_4 emissions during non-rice season. In the southwestern winter paddy field, the long-term flooding of paddy fields in winter resulted in strong soil reducibility, and the number of methanogenic bacteria could not be effectively controlled, resulting in much higher CH_4 emissions in the following rice-growing season than that of paddy fields which drain in winter (Kang *et al.*, 2002). The application of ammonium fertilizer in paddy fields will mainly affect CH_4 emissions from three levels of rice plant, microbial community, and biochemistry (Ma *et al.*, 2010; Zhou *et al.*, 2017). At the rice plant level, nitrogen fertilizer promotes root development, which increases underground rice biomass, root exudates, and provides more methanogenic substrates for methanogens, thereby promoting CH_4 emissions. At the microbial community level, nitrogen fertilizer promotes the growth and activity of methane-oxidizing bacteria. Thus, CH_4 is oxidized in a large amount, thereby reducing CH_4 emissions. At the biochemical level, NH_4^+ and CH_4 have a similar chemical structure, so, they compete to produce CH_4, which promotes CH_4 emission. The impact of urea application on CH_4 emissions from paddy fields depends on the relative strength of these three factors (Schimel, 2000). The application of organic fertilizer and straw returning to the paddy field will increase the CH_4 emissions. On the one hand, the easily decomposable organic carbon in organic fertilizer and straw can be used as a substrate for the production of CH_4 bacteria. On the other hand, the rapid decomposition of organic fertilizer under flooding conditions accelerates the decreasing of methane-oxidizing bacteria (Eh) of paddy fields, which is conducive to the growth of methanogenic bacteria and activity maintenance, thereby promoting CH_4 emissions in the paddy field (Quan *et al.*, 2017; Zhao *et al.*, 2020).

Different farming methods have different levels of disturbance to the soil, which will affect the physical and chemical properties and biological properties of the soil, such as oxidation–reduction potential, aeration, structure, moisture dynamics, activities of methanogenic bacteria, and methane-oxidizing bacteria and the growth of rice roots, which in turn affect CH_4 emissions (Cai *et al.*, 2003; Zhao *et al.*, 2016). Previous studies found that the main reasons for no tillage reduction of CH_4 emissions from rice fields included reduction of the content of soluble organic carbon in soil, increase of the total iron oxide content and soil bulk density, and increase of soil porosity, especially soil macropores (Feng *et al.*, 2013; Li *et al.*, 2013). Mangalassery *et al.* (2014) also stated that CH_4 emission reduction is related to the increase of the geometric porous structure in the soil under no tillage treatment. Therefore, improving soil conditions and increasing aeration of paddy soil can reduce the CH_4 emission of no tillage measures. Under the condition of returning straw to the field, CH_4 emission reduction effect of no-tillage measures has been reduced (Zhao *et al.*, 2016, 2020). This is mainly because the straw returned to the field provides more respiratory substrates, thus promoting the metabolism of soil methanogens and stimulating CH_4 emissions (Feng *et al.*, 2013; Liu *et al.*, 2014). The difference in land use is also the main factor causing the difference in CH_4 emissions. Paddy fields provide anaerobic conditions due to submerged irrigation, which promotes the process of soil respiration to produce CH_4, thereby increasing CH_4 emissions (Smith *et al.*, 2008).

3.4.3 *N_2O Emission Reduction*

N_2O is one of the important greenhouse gases, with the warming potential is 269 times of CO_2 at a 100-year scale. Agricultural production, especially nitrogen fertilizer input, is the main source of N_2O emissions, accounting for 60% of total anthropogenic N_2O total emissions (IPCC, 2013). N_2O is mainly produced by soil nitrification and denitrification, and soil temperature and humidity are important influencing factors (Mosier, 1998; Smith and Conen, 2004). Nitrogen fertilizer input is a major driving factor that affects N_2O in farmland ecosystems and is also an important parameter for estimating agricultural N_2O emissions (IPCC, 2013; Yue *et al.*, 2019). Many studies have focused on how to reduce farmland N_2O emission, but there is no consistent strategy at present.

There are many studies on intermittent irrigation, conservation tillage, no-tillage, returning straw to the field, applying biochar, and adding nitrification inhibitors, but with the environmental changes, the emission reduction effects often change accordingly (Zhao *et al.*, 2016, 2019).

The emission regulation of N_2O is relatively complicated, and the existing research results are not consistent. Studies have shown that, on the whole, no-tillage has no significant impact on N_2O emissions, but under the condition of returning straw to the field, no-tillage retention increases N_2O emissions in acid soil from paddy fields in the initial period of no-tillage within 5 years (Zhao *et al.*, 2016). Meta-analysis on different scales reported that the simultaneous use of no-tillage and straw returning to the field will increase N_2O emissions (Feng *et al.*, 2013; Mangalassery *et al.*, 2014), but VanKessel *et al.* (2013) stated that less or no-tillage will not significantly increase N_2O emissions. N_2O production is driven by soil nitrification and denitrification with the participation of soil microorganisms. This process is affected by soil aeration, water temperature, soil structure, compactness, organic matter distribution, and soil temperature and soil pH (Godde and Conrad, 2000; Cai *et al.*, 2003). Studies showed that returning straw to the field significantly increases N_2O emissions (Feng *et al.*, 2013; Liu *et al.*, 2014), because the straw returned to the field provides a respiratory substrate for microorganisms, stimulates microbial activity, accelerates oxygen depletion, promotes denitrification, and strengthens N_2O emission (Chen *et al.*, 2013; Zhao *et al.*, 2020). Adjusting the soil pH may be an important means related to N_2O emissions. The increase in soil organic matter input can increase soil pH and reduce acidic components, thereby changing the soil denitrification rate and N_2/N_2O emission ratio. In addition, soil pH cooperates with the soil redox electrolyte to further affect the decomposition process of organic matter. A meta-analysis study in the United States showed that compared with traditional farming measures, no-tillage in the central and eastern regions promotes N_2O emissions but reduces N_2O emissions in the central and western regions (Van Kessel *et al.*, 2013). This regional difference may be caused by different crop types, climatic conditions, and soil types. However, some meta-analysis results showed that the effect of no-tillage N_2O emissions under different climate or soil conditions is not significant (Li *et al.*, 2013; VanKessel *et al.*, 2013). Ussiri *et al.* (2009) pointed out that N_2O emission flux is highly correlated with rainfall, air, and soil temperature. Inconsistency of research results related to N_2O emissions is mainly due to differences in soil denitrification, soil water content, the number of substrates that can be effectively utilized by denitrifying bacteria, total soil nitrogen content, and soil temperature (Choudhary *et al.*, 2002; Mangalassery *et al.*, 2014). Global scale meta-analysis results showed that site condition, such as the time of indoor research, the C/N ratio of straw returning to the field, the amount of straw returning to the field, soil moisture content, soil structure, and clay content could significantly affect N_2O emissions (Chen *et al.*, 2013).

3.5 Development Strategies of CSA

3.5.1 *Enhancing the Application of Climate-Smart Irrigation in Response to Climate Change*

Optimal management of water and soil resources in agricultural production systems is an effective measure to adapt to climate change, but the extension coverage is not high. There are considerable obstacles to the adoption of these technologies. The main reason is that new adaptation technologies have led to increased labor/capital investment and return on investment is delayed. Farmers in the regions with greater climate risk could diversify agricultural production, labor, and income, thereby reducing vulnerability to extreme weather events. Previous studies showed that water management is key to adapting to climate change and improving agricultural resilience, but it requires a lot of public investment, and in the case of resource scarcity, this may be a serious problem. Irrigation plans after the initial investment may also result in smaller benefits than initially expected. In order to adopt climate-smart irrigation more widely, policymakers, researchers, and farmers urgently need to change their way of thinking and explore these opportunities in specific locations and specific conditions to facilitate local adaptation. The most important priority for combating climate change is irrigation, as it is an essential measure to protect farmers and others from climate risks. Taking Africa as an example, agriculture is highly dependent on rainfall, and huge differences of rainfall on the temporal and spatial scale have happened in recent years. Irrigation has been an essential point for stable and increased agricultural production. Irrigation not only alleviates the shortage of water resources, but also expands the sowing date of crops. Besides, it could increase the return on investment of fertilizers and other inputs. Irrigation is the key to crop production. After focusing on the construction of large-scale irrigation infrastructure for production development during a period of time, people are now paying more and more attention to irrigation projects to adapt to climate change, and these projects have been listed as a new priority issue. Improving local demand and appropriately adjusting irrigation management are used to change production methods, protect landscapes, and water resources, and make the most effective use of designed facilities. Operating mechanisms are reformed to better use existing infrastructure systems to increase water use and management efficiency. A variety of water use methods are considered to promote water-saving practices. Payments for Ecosystem Services (PES) plans are implemented to share part of the water use and better use the profits of forest growers and protectors to regenerate water resources in the upstream basin. Investments have been made in important disaster prevention infrastructure. Strengthened local management is undertaken on small and medium-sized irrigation infrastructure. Setting up a management organization is the key to managing water properly and improving adaptation efficiency. The cost of irrigation water and source of water resources are important for the sustainable development of the irrigation system. Technological innovation can play a major role in promoting sustainable water use. Regardless of how irrigation water enters the farmland, from furrow irrigation and gravity flow to sprinkler irrigation or drip irrigation, irrigation water can be used more effectively and is conducive to climate change. Drip irrigation is usually regarded as a water-saving technology, which is very helpful for saving resources and increasing water resource productivity (FAO, 2013).

3.5.2 *Integrating Climate Change into Agricultural Research and Technology Extension*

Incorporating the impact of climate change into agricultural research and extension activities can help improve the ability of agricultural production to adapt to climate change. Relevant case reports from FAO showed that (1) strengthening relevant research could help in devising management measures for specific climatic conditions and agricultural characteristics (e.g. crop variety selection plan, agricultural practices that coincident with simplified requirements) and adapted soil and water management measures for local agroecological conditions should be clarified and (2) continuous farmer training and extension plans can improve the resilience of farming families and ensure food security. Strengthening agricultural producers' access to meteorological data is very important and it should be given higher priority. At present, there is a gap between weather data and agricultural data, and there is almost no integration within them. Farmers from local to global, or any type of agricultural producers, lack the ability to use climate information. It is better for small farmers to have more knowledge about climate change, so education and joint actions between climate and agricultural technology service agencies are needed. In addition, lack of capacity is still a key limiting factor. Not only farmers lack the ability to acquire and absorb knowledge, but related knowledge institutions also lack the ability to deal with new and complex problems. Therefore, it is not only necessary to train and educate farmers, but also to train and educate the people who influence policies.

3.5.3 *Giving Crop Production a Priority to Cope with Climate Change*

The sustainable development of agriculture in the future requires a more comprehensive and systematic approach to analyze agricultural production systems, and research on crop production adapting to climate change should be given a higher priority. Public research institutions in developing countries are increasingly becoming aware that, given that the current working environment in the agricultural sector has changed, agriculture is facing water scarcity, lack of soil nutrients, climate change, agricultural energy supply, and loss of biodiversity. Due to emergence of new pests and diseases, the decentralization of farms, rural–urban migration, and new intellectual property rights and trade regulations, agricultural research programs must be transformed to fully utilize the potential of modern science, encourage technological innovation, and provide favorable policies and investment support. Priority must be given to key technologies such as genomics, molecular breeding, integrated technology of crop water and fertilizer, nanotechnology, secondary agriculture, and agricultural mechanization in future crop production research. We already have a lot of information to support better agricultural system research, but we need better coordination to obtain this information effectively. Considering the huge differences in biodiversity, scale, management systems, cultural diversity, and resource base, Kosura clarified the technical approach to establish representative types of agricultural systems.

3.5.4 *Strengthening Agricultural Technology Extensions*

Extension usually refers to a top-bottom and one-way technology transfer method, which has been replaced by the Agricultural Knowledge Information System (AKIS) for a long time, and subsequently replaced by agricultural consulting services. All experts agreed that it was absolutely necessary to shift from a top-down system to a system where knowledge flows in multiple directions. Based on current challenges, new promotion methods should be adopted to set up new promotion agencies. The basis of past agricultural revolutions was major disruptive innovations such as genetics, mechanization, or chemical inputs. The future agricultural transformation or the future agricultural revolution must evaluate different types of innovation. Combining disruptive innovation and technology with the expertise of farmers can make the best choice. Since the Green Revolution, there has been a competitive relationship between commodity-oriented promotion and agricultural system-oriented promotion, which to a certain extent is related to the changing roles of the public and private sectors. Research in India showed that only 6.5% of the obtained information by farmers came from public promotion, 20% from connections among farmers, and 20–29% from newspapers, radio, and television. With research and input delivered to the private sector, input and market access have become more and more important in a diversified agricultural production system. Private distributors have become an important source of information for farmers. At the same time, the new private department extension systems have become part of a growing value chain, and emerging input providers are learning to provide comprehensive services to farmers.

The role of public extension systems and governments in technology transfer needs to be redefined. With increasing emphasis on sustainable agriculture, the focus of extensions should be on the management of natural resources in the entire agricultural system, including water, soil, agriculture, forestry, and climate change. Generally speaking, the traditional extension system uses technical personnel who specialize in certain aspects of agriculture science, such as agronomy, plant pathology, soil science, plant breeding, animal husbandry, fishery, without using agricultural system methods to fully understand agriculture. Driven by supply, the role of public sector extension services in disseminating information has been weakened, exceeded the needs or expectations of farmers, and is unable to manage external factors on the farm. Therefore, it is necessary to establish appropriate incentive mechanisms in the extension system to promote higher quality services and better interaction and communication with farmers. Adaptive research and extension programs can provide effective connections between researchers and farmers. Innovative methods can be adopted to use information and communication technology, such as advisory services, thus providing information from private sector suppliers to farmers.

3.5.5 *Encouraging Development of Agricultural Insurance*

Insurance in agriculture is becoming more and more important, and subsidized insurance is an important way to encourage people to take to insurance. Index insurance implemented in Uganda and Zambia in Africa has been proved to be an effective tool for managing climate risk. Weather index-based insurance can enable farmers to obtain the benefits of risk control. Index insurance effectively promotes agricultural insurance through the establishment of effective management methods, mainly including training local insurance companies to engage in agricultural insurance business. It helps people understand the significance of keeping records and carrying out work as a group or through group ownership. In recent years, India has launched a program called Pradhan Mantri Fasal Bhima Yojana (PMFBY) (Prime Minister's Crop Insurance Scheme). This new solution deals with problems such as higher insurance premiums and insufficient insurance coverage. It also expands the definition of risk, including production loss, preventive sowing, and post-harvest losses. Farmers are now paying uniform insurance premiums based on the types of crops, and the government will fully build the bridge between the actual premiums and the farmers' payable rate. In order to improve the sustainability of public finances and policies, insurance as a risk management product invested by private financing institutions should be combined with the government's poverty reduction policies. Whether the government should support an index insurance plan for small household heads is a fiscal policy issue, that is, subsidizing a certain amount of insurance for the subsidy object is the most effective use of government resources or assistance provided by foreign institutions or non-government organizations (FAO, 2017).

3.5.6 *Developing Priorities of CSA*

Policy consistency is the highest priority for developing CSA. Climate change must be integrated and coordinated in the agricultural and non-agriculture sectors to realize better outcomes. An incentive mechanism should be established to encourage coordination and unification between the government and ministries and encourage many actors to adjust their behavior. In the decision-making and implementation work at the central and provincial levels, it is necessary to recognize that climate change activities need to be converged. A supporting evidence base must be established, and the need for trade-offs and compromises must be clearly recognized, which is effective for achieving effective coordination. To successfully implement climate change initiatives, it is necessary to rationalize/coordinate various government regulations, credit policies, subsidy programs, and land tenure laws and effectively incorporate these initiatives into departmental planning and budgets. It is also necessary to integrate different government departments with their local offices to deal with climate change issues at the landscape level, so as to be able to use community or participatory methods to effectively implement climate change adaptation plans and implementations at least cost. Reducing duplication and redundancy is another important aspect.

It is important to revise relevant policy plans and promote the participation of stakeholders and institutions. The required capabilities of institutions to participate in the climate change planning process vary from country to country, but in general, training and skills capital need to be improved through relevant training and technology development. Resources and financial capital can be mobilized through targeted projects to promote the development of CSA. Clear policy and management frameworks should be formulated to form political will. Public and private sector meetings and roundtable discussions also must be maintained regularity in order to ensure political will. They are critical resource mobilization strategies to the success of reform institutions, especially legislation institutions.

Strengthened interdisciplinary research will help to form more complete CSA programs. There are many institutions involved in the formulation and dissemination of information, so there must be a policy framework that encourages exchanges between different departments, ministries, private companies, and farmers' associations. There will be trade-offs and synergy between promoting productivity and environmental issues, so a favorable government environment is needed to deal with these issues in a reasonable way. Education and information are very important to facilitate this process, especially in developing countries.

It is necessary to incorporate climate change considerations into sector development plans. Vietnam is currently undergoing adjustments of agricultural structure, adjusting and implementing long-term plans, strategies, policies, organizational innovations, and improving public investment simultaneously within and outside the department and at all levels of managements. It is necessary to improve and strengthen communication and advocate for a change in the thoughts of managers from the central to local level. Evidence-based mechanisms and public support should also be part of effective management of natural resources. Both of them require good scientific information and research activities. Finally, forming a unified coordination system and an effective cross-departmental and regional coordination mechanism within a long-term action plan is the key to promoting effective integration.

References

Antle J M, Valdivia R O, Boote K J, *et al.* 2015. AgMIP's trans-disciplinary agricultural systems approach to regional integrated assessment of climate impact, vulnerability and adaptation. In C Rosenzweig, D Hillel (eds.) *Handbook of Climate Change and Agroecosystems: The Agricultural Model Intercomparison and Improvement Project Integrated Crop and Economic Assessments, Part 1.* London: Imperial College Press.

Barrett C B, Barnett B J, Carter M R, *et al.* 2007. Poverty Traps and Climate Risk: Limitations and Opportunities of Index-Based Risk Financing. *Built Environment eJournal.*

Bosse U, Frenzel P. 1997. Activity and distribution of methane-oxidizing bacteria in flooded rice soil microcosms and in rice plants (Oryza sativa L.). *Applied and Environmental Microbiology*, 63 (4): 1199–1207.

Bryan E, Deressa T T, Gbetibouo G A, *et al.* 2009. Adaptation to climate change in Ethiopia and South Africa: Options and constraints. *Environmental Science and Policy*, 12 (4): 413–426.

Cai Z C, Tsuruta H, M Gao. 2003. Options for mitigating methane emission from a permanently flooded rice field. *Global Change Biology*, 9 (1): 37–45.

Challinor A, Wheeler T, Craufurd P, *et al.* 2007. Adaptation of crops to climate change through genotypic responses to mean and extreme temperatures. *Agriculture, Ecosystems and Environment*, 119 (1): 190–204.

Chen H, Li X, Hu F, *et al.* 2013. Soil nitrous oxide emissions following crop residue addition: A meta-analysis. *Global Change Biology*, 19 (10): 2956–2964.

Chen C, Lei C, Deng A, *et al.* 2011. Will higher minimum temperatures increase corn production in Northeast China: An analysis of historical data over 1965–2008. *Agricultural and Forest Meteorology*, 151 (12): 1580–1588.

Chen H, Wang J, Huang J. 2014. Policy support, social capital, and farmers' adaptation to drought in China. *Global Environmental Change*, 24: 193–202.

Choudhary M A, Akramkhanov A, Saggar S. 2002. Nitrous oxide emissions from a New Zealand cropped soil: Tillage effects, spatial and seasonal variability. *Agriculture Ecosystems & Environment*, 93: 33–43.

Deressa T T, Hassan R M, Ringler C, *et al.* 2009. Determinants of farmers' choice of adaptation methods to climate change in the Nile Basin of Ethiopia. *Global Environmental Change*, 19 (2): 248–255.

Dinar A, Blankespoor B, Dinar S, *et al.* 2010. Does precipitation and runoff variability affect treaty cooperation between states sharing international bilateral rivers. *Ecological Economics*, 69 (12): 2568–2581.

Dobermann A, Simon B, Simon E C, *et al.* 2004. Adamchuk. Precision farming: Challenges and future directions. In *Proceedings of the 4th International Crop Science Congress*, 26 Sep -1 Oct 2004, Brisbane, Australia. Published on CDROM. Web site www.cropscience.org.au.

Easterling D R, Meehl G A, Parmesan C, *et al.* 2000. Climate extremes: Observations, modeling, and impacts. *Science*, 289 (5487): 2068–2074.

FAO. 2013. *Climate-smart Agriculture: Sourcebook*. Rome: FAO.

FAO. 2016. *Supporting Agricultural Extension Towards Climate-Smart Agriculture an Overview of Existing Tools*. Rome: Food and Agriculture Organization.

FAO. 2017. *Improving Climate Risk Transfer and Management for Climate-smart Agriculture: A Review of Existing Examples of Successful Index-based Insurance for Scaling Up*. Rome: Food and Agriculture Organization.

Feng J, Chen C, Zhang Y, *et al.* 2013. Impacts of cropping practices on yield-scaled greenhouse gas emissions from rice fields in China: A meta-analysis. *Agriculture, Ecosystems and Environment*, 164: 220–228.

Godde M, Conrad R. 2000. Influence of soil properties on the turnover of nitric oxide and nitrous oxide by nitrification and denitrification at constant temperature and moisture. *Biology and Fertility of Soils*, 32 (2): 120–128.

Howden S M, Soussana J F, Tubiello F N, *et al.* 2007. Adapting agriculture to climate change. *Proceedings of the National Academy of Sciences of the USA*, 104 (50): 19691–19696.

Huang J K, Wang Y J 2014. Financing sustainable agriculture under climate change. *Journal of Integrative Agriculture*, 13 (4): 698–712.

IPCC. 2001. *An Assessment of the IPCC. Synthesis Report — Summary for Policymakers*. Cambridge, UK and New York, USA: Cambridge University Press.

IPCC. 2013. Climate change 2013: The physical science basis. Contribution of Working Group I to the Fifth Assessment Report of the Intergovernmental Panel on Climate Change. Cambridge, UK and New York, NY, USA.

IPCC. 2014. Climate change 2014: Mitigation of climate change. In O Edenhofer, R Pichs-Madruga, Y Sokona, *et al.* (eds.) Contribution of Working Group III to the Fifth Assessment Report of the Intergovernmental Panel on Climate Change. Cambridge, UK and NY, USA, Cambridge University Press.

Kang G D, Cai Z C, Feng Z H. 2002. Importance of water regime during the non-rice growing period in winter in regional variation of CH_4 emissions from rice fields during following rice growing period in China. *Nutrient Cycling in Agroecosystems*, 64 (1–2): 95–100.

Kato E, Ringler C, Yesuf M. 2011. Soil and water conservation technologies: A buffer against production risk in the face of climate change Insights from the Nile basin in Ethiopia. *Agricultural Economics*, 42 (5): 593–604.

Khanna M, Zilberman D. 1997. Incentives, precision technology and environmental protection. *Ecological Economics*, 23 (1): 25–43.

Lal R. 2011. Sequestering carbon in soils of agro-ecosystems. *Food Policy*, 36: 33–39.

Larson D F, Dinar A, Blankespoor B. 2012. Aligning Climate Change Mitigation and Agricultural Policies in Eastern Europe and Central Asia. *Asia and the World Economy* (69–151). World Bank, Washington, DC.

Larson D F A, Dinar B B. 2015. Aligning climate change mitigation and agricultural policies in ECA. In J Whalley (eds.) *Asia and the World Economy* (pp. 69–151).

Li C, Zhang Z, Guo L, *et al.* 2013. Emissions of CH_4 and CO_2 from double rice cropping systems under varying tillage and seeding methods. *Atmospheric Environment*, 80: 438–444.

Liu C, Lu M, Cui J, *et al.* 2014. Effects of straw carbon input on carbon dynamics in agricultural soils: A meta-analysis. *Global Change Biology*, 20 (5): 1366–1381.

Lobell D B, Bänziger M, Magorokosho C, *et al.* 2011. Nonlinear heat effects on African maize as evidenced by historical yield trials. *Nature Climate Change*, 1 (1): 42–45.

Lu L, Reardon T, Zilberman D. 2016. Supply Chain Design and Adoption of Indivisible Technology. *American Journal of Agricultural Economics*, 98:1419–1431.

Lu L, Thomas R, David Z. 2015. Supply chain design and adoption of indivisible technology. In *Presentation at Allied Social Sciences Association Annual Meeting*.

Ma J, Xu H, Cai Z C. 2010. Effect of fertilization on methane emissions from rice fields. *Soil*, 2: 153–163.

Mangalassery S, Sjogersten S, Sparkes D L, *et al.* 2014. To what extent can zero tillage lead to a reduction in greenhouse gas emissions from temperate soils? *Scientific Report*, 4: 4586.

Meng Q, Hou P, Lobell D B, *et al.* 2013. The benefits of recent warming for maize production in high latitude China. *Climatic Change*, 122 (1–2): 341–349.

Mosier A R 1998. Soil processes and global change. *Biology and Fertility of Soils*, 27 (3): 221–229.

Olesen J E, Trnka M, Kersebaum K C, *et al.* 2011. Impacts and adaptation of European crop production systems to climate change. *European Journal of Agronomy*, 34 (2): 96–112.

Olmstead A L, Rhode P W. 2011. Adapting North American wheat production to climatic challenges, 1839–2009. *Proceedings of the National Academy of Sciences*, 108 (2): 480–485.

Parfitt J, Mark B, Sarah M. 2010. Food waste within food supply chains: Quantification and potential for change to 2050. *Philosophical Transactions of the Royal Society of London B: Biological Sciences*, 365 (1554): 3065–3081.

Parker D, David Z. 1996. The use of information services: The case of CIMIS. *Agribusiness*, 12 (3): 209–218.

Poteete A R, Janssen A, Elinor O. 2010. *Working Together: Collective Action, the Commons, and Multiple Methods in Practice*. Princeton, NJ: Princeton University Press.

Quan X F, Yan X Y, Wang S W. 2017. Effects of long-term application of organic materials on the ecosystem services of paddy fields. *Journal of Agro-Environmental Sciences*, 7: 1406–1415.

Reardon T, Timmer P C. 2012. The economics of the food system revolution. *Annual Review of Resource Economics*, 4 (1): 225–264.

Reidsma P, Ewert F, Lansink A O, *et al.* 2010. Adaptation to climate change and climate variability in European agriculture: The importance of farm level responses. *European Journal of Agronomy*, 32 (1): 91–102.

Reidsma P, Ewert F, Lansink A O, *et al.* 2009. Vulnerability and adaptation of European farmers: A multi-level analysis of yield and income responses to climate variability. *Regional Environmental Change*, 9 (1): 25–40.

Reidsma P, Ewert F. 2007. Oude-Lansink, analysis of farm performance in Europe under different climatic and management conditions to improve understanding of adaptive capacity. *Climatic Change*, 84 (3–4): 403–422.

Rosenzweig C, Jones J W, Hatfield J L, *et al.* 2015. In Handbook of Climate Change and Agroecosystems: The Agricultural Model Intercomparison and Improvement Project (AgMIP) Integrated Crop and Economic Assessments, Part 1. C. Rosenzweig and D. Hillel, Eds., *ICP Series on Climate Change*

Impacts, Adaptation, and Mitigation, vol. 3, Imperial College Press, pp. 331–386.

Schimel J. 2000. Global change-Rice, microbes and methane. *Nature*, 403 (6768): 375.

Schlenker W, Roberts M. 2009. Nonlinear temperature effects indicate severe damages to U. S. crop yields under climate change. *Proceedings of the National Academy of Sciences of the USA*, 106 (37): 15594–15598.

Smit B, Skinner M W. 2002. Adaptation options in agriculture to climate change: A typology. *Mitigation and Adaptation Strategies for Global Change*, 7 (1): 85–114.

Smit B, Wandel J. 2006. Adaptation, adaptive capacity and vulnerability. *Global Environmental Change*, 16 (3): 282–292.

Smith K A, Conen F. 2004. Impacts of land management on fluxes of trace greenhouse gases. *Soil Use and Management*, 20: 255–263.

Smith P, Martino D, Cai Z, *et al.* 2008. Greenhouse gas mitigation in agriculture. *Philosophical Transactions of the Royal Society B-Biological Sciences*, 363 (1492): 789–813.

Stringer L C, Dyer J C, Reed M S, *et al.* 2009. Adaptations to climate change, drought and desertification: Local insights to enhance policy in southern Africa. *Environmental Science and Policy*, 12 (7): 748–765.

Tao F, Zhang Z. 2010. Adaptation of maize production to climate change in North China Plain: Quantify the relative contributions of adaptation options. *European Journal of Agronomy*, 3 (2): 103–116.

Tao F, Yokozawa M, Xu Y, *et al.* 2006. Climate changes and trends in phenology and yields of field crops in China, 1981–2000. *Agricultural and Forest Meteorology*, 138 (1–4): 82–92.

Thornton P K, Jones P G, Alagarswamy G, *et al.* 2010. Adapting to climate change: Agricultural system and household impacts in East Africa. *Agricultural Systems*, 103 (2): 73–82.

Tol Richard S J. 2009. The economic effects of climate change. *The Journal of Economic Perspectives*, 23(2): 29–51.

Tubiello F N, Soussana J F, Howden S M. 2007. Crop and pasture response to climate change. *Proceedings of the National Academy of Sciences*, 104 (50): 19686–19690.

Ussiri D A N, Lal R, Jarecki M K. 2009. Nitrous oxide and methane emissions from long-term tillage under a continuous corn cropping system in Ohio. *Soil & Tillage Research*, 104 (2): 247–255.

Van Kessel C, Venterea R, Six J, *et al.* 2013. Climate, duration, and N placement determine N_2O emissions in reduced tillage systems: A meta-analysis. *Global Change Biology*, 19 (1): 33–44.

Wang J, Huang J, Zhang L, *et al.* 2010b. Why is China's Blue Revolution so "Blue"? The determinants of conservation tillage in China. *Journal of Soil and Water Conservation*, 65 (2): 113–129.

Wang J, Mendelsohn R, Dinar A, *et al.* 2010a. How Chinese farmers change crop choice to adapt to climate change. *Climate Change Economics*, 1 (03): 167–185.

Waterfield G, Zilberman D. 2012. Pest management in food systems: An economic perspective. *Annual Review of Environment and Resources*, 37: 223–245.

Ye R Z, Horwath W R. 2017. Influence of rice straw on priming of soil C for dissolved organic C and CH_4 production. *Plant and Soil*, 417: 231–241.

Yue Q, Wu H J, Sun F, *et al.* 2019. Deriving emission factors and estimating direct nitrous oxide emissions for crop cultivation in China. *Environmental Science & Technology*, 53: 10246–10257.

Zhao X, Liu B Y, Liu S L, *et al.* 2020. Sustaining crop production in China's cropland by crop residue retention: A meta-analysis. *Land Degradation & Development*, 31: 694–709.

Zhao X, Pu C, Ma S T, *et al.* 2019. Management-induced greenhouse gases emission mitigation in global rice production. *Science of the Total Environment*, 649: 1299–1306.

Zhao X, Liu S L, Pu C, *et al.* 2016. Methane and nitrous oxide emissions under no-till farming in China: A meta-analysis. *Global Change Biology*, 22: 1372–1384.

Zhou M, Zhu B, Wang X, *et al.* 2017. Long-term field measurements of annual methane and nitrous oxide emissions from a Chinese subtropical wheat-rice rotation system. *Soil Biology and Biochemistry*, 115: 21–34.

Chapter 4

Monitoring, Assessment and Extension of Climate-Smart Agriculture Projects

Abstract

Monitoring, assessment and extension are important links in the development of climate-smart agriculture (CSA). To facilitate these, the first task is to build a complete set of CSA monitoring and assessment systems and clarify their scopes, purposes and frameworks. Second, the assessment of CSA is completed through the following steps: designing assessment, climate impact assessment and CSA option assessment. Finally, the methodology and practice of CSA extension are improved with respect to various aspects, including agrometeorological forecast and farmers' field school. Extension and technical innovation of CSA were further strengthened through the application of digital agriculture, early warning system, CSA and farmers' organization, etc. We should make efforts together in monitoring, assessment and extension to coordinate and jointly promote the sustainable development of CSA.

4.1 Monitoring and Assessment Frameworks and Goals of CSA Projects

The implementation of CSA requires effective monitoring and assessment methods. The agriculture and climate change departments work with multilateral development banks to promote the implementation of CSA. For example, the World Bank Group (WBG) has invested billions of dollars in CSA projects. At the same time, the national projects need to meet certain conditions of donors to prove the rationality of funding. Climate financing institutions such as the Global Environment Facility (GEF), Green Climate Fund (GCF) and United Nations Framework Convention on Climate Change (UNFCCC) have adopted the development of a monitoring and assessment framework to assess the project's contribution to reducing climate vulnerability and enhancing climate resilience. Since 2015, governments have pledged to reach three global agreements: the 2030 Agenda for Sustainable Development, the Sendai Framework for Disaster Risk Reduction (2015–2030), and the Paris Agreement of the UNFCCC. The "Paris Agreement" specifically requires countries to report on the progress of climate change mitigation and adaptation goals and plans, and provide corresponding information for the overall progress of the upcoming agreement. These CSA projects help countries fulfill their commitments to the "Paris Agreement" and sustainable development goals. To encourage major donors to maintain consistency with the indicators, monitoring and assessment frameworks of the three global agreements are very important for CSA projects, as these help reduce the reporting burden of countries and avoid duplication. Monitoring these goals is very complicated, especially monitoring adaptation and resilience; while tracking and measuring the funds flowing into the CSA projects, it is also necessary to reduce the burden of reporting due to the different requirements of each agreement. Therefore, it is very important to build reasonable monitoring and assessment systems.

4.1.1 *Assessment for CSA Projects Designs*

CSA is a region-specific system project, and there is no universal approach suitable for all situations. It is necessary to formulate corresponding policies and programs on the basis of a detailed strategic assessment of the benefits and losses under various social, economic and environmental conditions. This assessment can verify the local climate conditions and their impact on agriculture, food security and livelihoods, and then expected future changes. In addition, this assessment can also determine whether certain measures are beneficial to mitigating climate change under certain circumstances. It is worth noting that good CSA measures may be different from traditional agricultural development and natural resource management.

An assessment of policy and project design can verify the impacts of climate change on the agricultural sector, food security and livelihoods. The climate impact assessments verify the impact characteristics of climate change and the most vulnerable locations and environments that require adaptation actions. Climate impact assessment is a kind of strategic planning, usually carried out during the conceptualization phase of a project, and is also used to directly provide information for policies. The main contents of climate impact assessment include the following: climate change, the impact of climate change on agricultural production, livelihood vulnerability, food security, and climate change adaptation. Analyzing the relationship between climate and agriculture can combine relevant current and past data to predict possible future climate changes and further infer the potential impact on agricultural production, as well as the corresponding description of vulnerability. On this basis, a baseline forecast is made in accordance with the results of the assessment.

There is great significance to systematically assess weather changes in agricultural and commercial operations as well as the policy environment, and it is conducive to achieving CSA goals such as adaptation, mitigation, food security and development. Generally, the assessment of CSA alternatives is carried out in the project preparation stage, and the corresponding personnel determines the project and program baseline, so as to finally determine the implementation plan of it. Based on historical, current and predicted climate impacts on vulnerability analysis to livelihoods and food security, we can simulate and select more suitable strategies for adapting to climate change, and then explore effective adaptation plans for climate change. Mitigation of climate change can be achieved in two ways: (1) reducing greenhouse gas emissions; (2) storing more carbon in biomass and soil. Evaluating the mitigation potential of different measures is very important for mitigating the impact of climate change. At the national level, monitoring and evaluating the impact of carbon storage are of great significance to international agreements on climate change mitigation. CSA should also meet the needs of broader food security and development goals. The best intervention measure for CSA is to effectively promote adaptation and mitigation of climate change without compromising food security and development goals. Ideally, priority should be given to options that include as many CSA goals as possible. The assessments of policies and project design are carried out before the intervention, and some of them need to run through the entire project cycle. These can cover climate, biophysical and socioeconomic aspects.

4.1.1.1 *Climate Impact Assessment*

Climate impact assessment concretizes historical, current or future changes, and based on this, it establishes the relationship between productivity and climate change in the agriculture, forestry, and fisheries sectors. Climate impact assessment can visually demonstrate the vulnerability of different stakeholders to climate change and the potential impact of climate change on agriculture. Among them, the stakeholders targeted by the vulnerability assessment include male and female small farmers, landless labor, commercial farmers, and people working in the value chain. In climate impact assessment, climate impact refers to the impact of climate change on natural and human systems. In CSA, climate impacts are reflected in different scales, such as landscapes, ecosystems, watersheds, infrastructure, farms, agricultural production, and markets. The climate impact assessment considers the degree of exposure and sensitivity to exposure. Vulnerability is a combination of potential impact and adaptive capacity (Carter *et al.*, 2007). Adaptability refers to the system's ability to avoid potential damage, take advantage of opportunities, and deal with the consequences of damage. Resilience refers to the ability or effect of a system and its components to anticipate, absorb, accommodate or recover from a dangerous event (IPCC, 2012). Adaptive ability is the ability of a given system to affect resilience (Walker *et al.*, 2004; Folke, 2006; Engle, 2011). Climate impact assessment can provide important references for science, policy and the public. An effective climate impact assessment must take into account a wide range of stakeholders, verify the process, interpret the results, and translate it into support for climate change adaptation and mitigation options.

4.1.1.2 *CSA Options Assessment*

After the climate impact assessment, the CSA options assessment can evaluate the extent to which different CSA measures can achieve the goals of increasing productivity, enhancing climate change adaptation and mitigation capabilities and improving food security, and clarify the expected impact of climate change. This helps relevant personnel to identify effective CSA options and work together to achieve multiple goals. In addition, CSA projects can be formulated to implement relevant effective measures; ideally, CSA strategies should be reviewed and updated regularly.

4.1.2 *Monitoring and Evaluation for CSA Programs and Projects*

Monitoring and evaluation are essential to ensure that CSA-related measures are properly implemented and achieve expected results. The monitoring and evaluation framework is designed after the evaluation of climate change and the formulation of CSA programs and detailed project plans. When monitoring and evaluating CSA projects, the baseline prediction of climate conditions and the expected goal of the policy and project design evaluation are the starting and end points, respectively. Monitoring and evaluation usually need to extract indicators from the design and evaluation of policies and projects, and describe the baseline conditions of the project based on the original data collected by analysis. In addition, CSA projects can also give priority to using relevant information from the evaluation.

In the project preparation stage, the monitoring and evaluation of the project are started in the interaction between evaluation, monitoring and evaluation activities. Through the development of detailed and regular plans, the related processes are closely linked, and the project and program baselines are guided based on the evaluation framework. After the project is evaluated, detailed indicators and baselines need to be determined, and beneficiaries and corresponding intervention measures should be clearly designated. Throughout the implementation of the project, the company's operations, resource use and outputs are monitored. In the middle and end of the project, based on the baseline situation and initial expected results, an assessment of the impact of CSA-related measures on socio-economic, environmental and livelihood indicators is carried out. Whether in a pilot or a project, supervisors must continuously monitor the effects of climate-smart agricultural measures and problems during the implementation process. This monitoring will verify whether the measures are implemented efficiently and meet the CSA goals and project goals. It also helps adjust activities in a timely manner in response to uncertainties that may occur. During the project cycle, monitoring and evaluation are conducive to the implementation of responsibilities and the reasonable allocation of resources. Good monitoring and evaluation can help improve the design of future climate-smart agricultural interventions and stakeholder decision-making, and achieve national mitigation goals. For example, under the framework of the UNFCCC, detailed monitoring of greenhouse gas emissions has become an important part of the measurement requirements.

4.1.3 *Baselines and Baseline Projections*

CSA evaluation needs to establish a baseline or predict the impact of specific policies based on relevant basic conditions. This method is conducive to timely and effective evaluation of the possible gains of the project or plan and the monitoring results and impacts. The baseline for climate impact evaluation can be formulated based on the predicted future climate change, related changes in agricultural output and other relevant information. This forecasting information needs to be carried out without planning or project intervention, mainly including climate impact, agriculture, food security and vulnerability status. This "no intervention" situation helps to formulate broader policies and programs to assess the long-term impact of CSA-related interventions.

A detailed evaluation of CSA options helps to clarify the project and program baselines without intervention as well as the changes in related indicators used to define the implementation effects of CSA projects after the intervention. At the end of the project cycle, the impact of climate-smart agricultural interventions will be evaluated on these projects and program baselines. During the implementation period, the intermediate results and the progress made in the results are constantly monitored.

As climatic conditions evolve during the project and program cycles and new information about the impact of climate change and the vulnerability to climate change emerges, it may be necessary to periodically revise baselines and forecasts. It needs to be designed according to changing climatic conditions. Since the carbon balance of the ecosystem is dynamic, it may change over time in the absence of mitigation interventions. The project manager may need to adjust the intervention measures of CSA based on the revised baseline forecast in the mid-term of the project.

For shorter CSA projects, the baseline changes may be biased (GIZ, 2011a). The use of the "control group" in the impact evaluation should allow for some baseline variability and changes in other factors, such as the market and the broader economy. For projects and plans over 5 years, monitoring and evaluation should be based on "mobile" or the latest basic projects as well as baselines for typical projects and plans. Therefore, baseline forecasts can be used to deal with longer-term climate change adaptation and mitigation measures.

4.2 Steps to Assess CSA Projects

4.2.1 *Designing Assessments*

Obtaining useful information through literature review and systematically sorting out relevant information about countries and regions are the first steps in planning CSA actions. The literature review identifies gaps by collecting and analyzing additional data and supplements them with custom assessments. Information about climate change and its effects on agriculture is available in large amounts on a global and regional scale. However, data at the national and sub-national levels are scarce, but they can be obtained from the following sources: the Intergovernmental Panel on Climate Change (IPCC) Fourth Assessment Reports, National Communications to the UNFCCC, UNFCCC National Adaptation Programmes of Action (NAPAs), academic papers, non-peer-reviewed reports and related research institutions.

When designing an assessment, it is necessary to clarify the information needs of CSA practitioners. This is because the impact and environmental potential assessments of CSA led by scientists and economists often ignore the opinions of practical users. Different stakeholders play different roles in assessment. Assessment of the role and ability of stakeholders is one of the most basic assessment projects, and it is also the key to successful assessment of policies and projects. The National Climate Change Office (NCCD), the Ministry of Agriculture and other relevant departments have determined the CSA goals and implementation paths of various countries. Practitioners of CSA are local government officials, extension workers or local farmers. Scientific assessment needs to consider the characteristics of stakeholders and their participation in designing assessments.

The main steps for designing assessment can be summarized as follows: (1) literature review; (2) identification of stakeholders; (3) assessment of information needs of stakeholders; (4) evaluation of role and capacity of stakeholders; (5) designing assessments, including objective stakeholders of CSA on agreements.

4.2.2 *Climate Impact Assessment*

4.2.2.1 *Climate Change*

Forecasting climate change and its impacts is one of the core contents of CSA. Meteorological observation data are obtained from thousands of land meteorological stations around the world, as well as some ships, radiosondes, aircraft, and satellites. Some of them are shared by the international communities, and some are owned by corresponding countries. The availability of data is the focus of evaluating the impact of climate change. The availability and quality of data will vary significantly from location, country, climate variables, and time frequency. In addition, the observed climate data can also be verified and supplemented with local knowledge of climate data. Future climate data are usually output from models, and global scale data can be converted to finer spatial scales through downscaling. For example, the Global Environment Facility (GEF) is committed to the protection and continuous management of natural resources, and the implementation of its integrated natural resource management project in Guinea and neighboring countries has played a demonstrative effect in climate change prediction. In the first phase of the project, it is important to reduce the vulnerability of rural communities to climate change. The project requires a clear understanding of the current impact of climate change to ensure that the corresponding measures have a positive impact on the resilience of climate change. Therefore, it is necessary to assess the impact of current climate-related disasters on the livelihoods of different agroecological areas and complete the climate risk evaluation. During the community consultations, farmers reported that the climate situation has undergone tremendous changes in the past decade (increased frequency of droughts, extreme high temperatures, delayed rainy seasons, and the scarcity and unpredictability of rainfall). These views on the time, intensity and frequency of climate change are consistent with the actual performance of scientific data and have a significant impact on farmers. Discussing and analyzing the views of local people on the connection between current and potential coping strategies, understanding how farmers deal with climate risks, and identifying resources sensitive to climate risks are the keys to implementing these strategies and helping in the integration of environmental protection wisdom into project activities and practice. The originally planned activities need to be adjusted according to the project approach because these activities may affect the availability of key livelihood resources and the access of local communities to these resources. The purpose of the revision is to adapt the project activities to the current climate conditions and help deal with the increasingly serious climate problems.

4.2.2.2 *The Impacts of Climate Changes on Agriculture*

Analyzing the impacts of climate change on agricultural production requires full consideration of factors, such as agricultural input, food demand, transportation, distribution channels, and the process of agricultural production activities. The impact assessment model is usually a highly specialized physical model or economic model, so experts in these fields are needed to give professional suggestions and assistance. Aquacrop, for example, is an FAO model crop to simulate yields of major crops' response to water. The Modeling System for Agricultural Impacts of Climate Change (MOSAICC) is an integrated package of tools for facilitating an interdisciplinary assessment of the impacts of climate change on agriculture.

The assessment of climate impacts on agricultural productivity generally takes a top-down approach, taking into account the impact of current and past climate change on CSA, local perceptions of climate change, long-term weather and the collection of agricultural historical data, all of which need to be concluded. Global climate models can not only provide future climate predictions based on socio-economic and emission scenarios but also reduce their scale by using appropriate methods. The calibrated model can be used to simulate the impact of future climate change on agriculture. The "United Nations Framework Convention on Climate Change (UNFCCC)" provides a review of existing agricultural models, including the following: agroclimatic index, geographic information systems (GIS), statistical models, process-based crop models, and economic models. The impacts of climate change on crop production include the following: changes in temperature and precipitation, variations of growing season lengths, carbon dioxide effects, outbreaks of pests and diseases, and the effectiveness of irrigation water. Changes in agricultural production will affect food production and have repercussions throughout the national economy.

4.2.2.3 *Vulnerability Assessment*

The impacts of climate change on agricultural productivity and the agricultural sector will affect household income and food security to varying degrees. The vulnerability of livelihoods depends on the ability of the local community's production systems if these systems could prevent the loss of agricultural income, provide self-sufficient production or supply food to the market to reduce adverse effects. Vulnerability assessment describes and identifies areas, families or groups with particularly low livelihood resilience, which helps community planning to prioritize their interests and make vulnerable communities the primary target of action. Vulnerability assessments also provide a basis for development and increase the relationship between livelihoods and climate change.

On the one hand, the potential impact of climate change identified in the previous assessment can be used to assess the vulnerability of the system through the assessment of the adaptive capacity of the system. On the other hand, the bottom-up approach focuses more on collecting different indicators that can describe the vulnerability of CSA and related sectors to various risks (including climate change), relevant indicators also include socioeconomic resources, technology, infrastructure, information and skills, institutions, biophysical conditions and universality (Dessai and Hulme, 2004). Climate change and changes in socio-economic, political and institutional structures can threaten the society and environment. Social and environmental conditions indicate their adaptability, ability, and vulnerability to potential threats.

4.2.3 *CSA Option Assessment*

4.2.3.1 *Adaptation*

The assessment of the adaptation effect of CSA options is an extension of climate impact assessment. After understanding the potential impacts of climate change and vulnerability, people could determine the best practices that are appropriate to local conditions, which helps select suitable and feasible adaptation options. For example, process models can be used to optimize water and fertilizer management to increase yields, and economic models can simulate the impact of fertilizer subsidies on productivity, market prices, and farm income. Screening analysis is carried out around the evaluators answering "yes" "or" "no" questions about the options. Those options with the most "yes" can be given the highest priority or further use quantitative analysis methods for assessment. For example, in the cost–benefit analysis, the relative costs of different adaptation options that achieve similar results are compared (UNFCCC, 2010b). There is no significance for local communities to fully participate. In a bottom-up approach, taking into account the local climate, socio-economic and environmental conditions (based on community adaptation), local male and female farmers discuss and decide that they are willing to adopt the best climate-smart agricultural interventions. This approach is conducive to linking local traditional knowledge with scientific knowledge. Participatory assessments also provide an opportunity for vulnerable groups to determine the unintended consequences of climate-smart agricultural interventions and how to address these consequences. When the comparative advantages of different adaptation options are not clear, the cost and benefit of adaptation measures can be evaluated through economic cost–benefit analysis or non-economic assessment methods (World Bank, 2009c). In addition, CSA should also provide an overall strategic framework to achieve the sustainability of agricultural producers and food security, and then add food security assessment criteria to the vulnerability and adaptation assessment models.

4.2.3.2 *Mitigation*

An assessment of mitigation benefits quantifies their mitigation potentials by simulating the dynamics of greenhouse gas emissions and carbon sequestration of CSA options. The assessments of mitigation potential typically assume a linear relation between the intensity of the mitigation activity and the estimated emission of a given emission and removal activity through an emission factor. The Ex-Ante Carbon-balance Tool (EX-ACT) and Marginal Abatement Cost Curves (FAO, 2012a) are some of the tools that facilitate the calculation of mitigation potentials for CSA projects. Other tools such as the Agriculture and Land Use National Greenhouse Gas Inventory (ALU) Software (Colorado State University, 2013) and the Carbon Benefits Project (CBP) tool (UNEP, 2013) are not specific to any region and can be applied in any location. A life cycle assessment (LCA) approach may be necessary to estimate the GHG emissions throughout the life cycle of a product, including the production, transportation and distribution of fertilizers and pesticides, product transportation, processing, packaging, and product distribution to retailers. The LCA is widely accepted in agriculture and industry. It is a method for evaluating the environmental impacts, production and emission enhancement processes during the product cycle (FAO, 2010a, 2012b).

4.3 Indicators and Methods for Evaluating CSA Projects

4.3.1 *Framework Design and Indicator Selection*

When formulating the basic framework for the implementation of CSA, it is necessary to fully consider monitoring and evaluation in the design process and target formulation. The sharing process of establishing goals and determining indicators among CSA stakeholders is the key to obtaining feedback, learning, and strategy formulation. They focus on climate change and agriculture, and their results and impacts are closely related to evaluation. Capacity development, organizational change, infrastructure and policy support are specific areas of high intervention, which are in line with higher overall planning and supervision guidance. The project and program framework helps to describe the expected outputs and results of stakeholder participation.

It is important to distinguish between process-oriented goals and result-oriented goals. An understanding of the underlying processes is critical to the implementation of CSA, but these processes are difficult to measure, so it is often overlooked. The implementation of CSA cannot be carried out strictly in a linear manner. With the rapid changes in the environment and the ever-increasing adaptability, changes in the process and participation must also be measured, for example, understanding the causes of behavior changes is worth monitoring and evaluating (Villanueva, 2010). In this field, we can learn from related work in the broader field of agricultural development and other disciplines (FAO, 2012d).

The evaluation framework developed by the Canadian International Development Research Center (CIDRC) within the scope of its research has been adopted by a series of programs to promote institutional reforms, and it has played an important role in describing the expected results between the partners and stakeholders involved in different projects. Result mapping is suitable for monitoring system changes, capturing capacity changes and the resulting service delivery. In order to measure the progress and achievements of the project, it is necessary to adopt appropriate indicators, use appropriate goals and means of each result, which is the core content of the project monitoring and evaluation framework. Indicators are an important part of evaluating corresponding measures. Indicators should be as simple, measurable, attributable, reliable and time-bound as possible. The indicator scope of CSA includes the adaptation and mitigation of the agricultural sector to climate change. In addition, intervention measures should help increase productivity, improve the ability to adapt to climate risks, reduce greenhouse gas emissions, increase greenhouse gas absorption, and increase the realization of national food security and development goals.

4.3.2 *Project Monitoring and Evaluation Methods*

The monitoring and evaluation indicators of CSA projects include technology adoption, land use changes, household livelihoods and institutional changes, involving qualitative and quantitative data analysis. Monitoring requires not only an internal integrated system to track financial transactions, expected output, activity goals and achievements but also feedback and learning to be incorporated into program and project management. Monitoring systems increasingly track and manage data through computerized management information systems. The key to successful monitoring and evaluation is to internalize their importance in planning and decision-making, reflecting the importance of management and other stakeholders in CSA interventions. Monitoring and evaluation tasks are often used to provide reports to governments or donors.

4.4 Extension Experience of CSA

4.4.1 *Methodology of Extension*

However, due to the lack of timely sharing of information and knowledge related to CSA, its acceptance is still limited. Although agricultural consulting services can bring information directly to the end users, many developing countries have insufficient service capabilities due to chronic shortages of manpower, limited business funds, and weak connections with other research participants. Data from Africa showed that each extension worker could serve 950 farmers in Kenya, 2,500 in Uganda, and 3,420 in Nigeria (Sones *et al.*, 2015). This situation has led to poor performance, limited coverage and limited impact of the extension system and is the main challenge for the implementation of CSA. These systemic constraints may not be conducive to the smooth implementation of adaptation strategies to quickly respond to the changing climate environment, thereby posing a threat to the agricultural economy. Extension services have traditionally been regarded as a mechanism for putting research-based knowledge into practical use, with a focus on increasing agricultural production. The Global Forum for Rural Advisory Services (GFRAS) reported that declining water availability, increasing soil degradation, and new global challenges such as climate and market changes and uncertainties will lead to tremendous changes in the role of today's extension systems in 2012. Responding to these global challenges requires the generation, adaptation and use of new knowledge and the interaction and support of a wide range of organizations. These new challenges also mean that the expansion of the system needs to deal with a variety of goals that go far beyond the transfer of new technologies, such as linking domestic and international markets more effectively and responsibly, and reducing the vulnerability of the rural poor, so as to increase their voice, promote environmental protection, establish connections between farmers and other institutions, and promote the development of farmers groups (Davis, 2009; GFRAS, 2012). The extension comes "in many sizes and shapes" (FAO, 1998), and a distinction between the extension approaches as such (e.g. participatory training approach, training and visit approach) or the main underlying principles of the advice (e.g. organic production, integrated production) is not absolute. All extension systems are facing the challenge of climate change. It is necessary not only to shift from focusing on food security to considering comprehensive factors but also to consider the differences between the three components of CSA. Synergy and trade-offs require substantial investment to develop extension and farmer-level knowledge and capabilities. The climate-smart extension system is a highly efficient and profitable tool and plays an important role in combating climate change. Agricultural extension methods mainly include the aspects discussed in the following sections.

4.4.1.1 *Agrometeorological Forecast*

Agrometeorological forecast is the basic element of the implementation of CSA. The meteorological and hydrological departments of various countries provide farmers with a series of weather information such as real-time weather data and short-term, medium-term and long-term weather forecasts, which are used to provide scientific support for the adjustment of agricultural production and promote the implementation of CSA. Agricultural climate consulting promotes technological innovation by building early warning and decision support systems, such as adopting advanced biotechnology in the field of genetics and adopting new technologies such as farming, irrigation, weeding, plant protection, soil planting and protection in the field of agricultural technology to promote agricultural production adaptation climate change. Climate services help optimize farm management, such as reducing temperature by using mulch, reducing evaporation, preventing soil erosion, improving soil fertility, planting trees for wind protection, preventing/reducing sand flow, microclimate modification, and building a good irrigation system to cope with drought, etc. The dissemination of agrometeorological information and tools is related to stakeholders in various countries to varying degrees. Some countries jointly create climate services for agriculture through dialogue between farmers and scientists. For example, farmers' production requires the development of appropriate meteorological service tools, such as daily rainfall measurements on farms, daily agricultural ecological observations, and their impacts on yield/quality and seasonal rainfall scenarios. Other weather services help increase farmers' adaptive potential, such as warnings of extreme times, such as drought, floods, heat waves, and heavy rainfall. Wireless networks, radio and broadcasting, face-to-face meetings and group meetings as well as dialogues, seminars and technical meetings are the main methods of disseminating agricultural meteorological information. These methods can provide end users (including farmers, actors of the value chain, and technical and support services) with appropriate weather and climate information suitable for specific needs and support users to carry out activities under various CSA goals.

The National Meteorological and Hydrological Services (NMHSs) and other international research organizations keep at the forefront of the application of agricultural meteorological tools. Climate services in different regions of China have been proven could enhance crop production's adaptation to adverse events. CAgM METAGRI projects conducted by World Meteorological Organization (WMO) also showed that farmers have the ability to determine when to plant seeds based on the rainfall data recorded by their self-made rain gauges. These methods can reduce energy demand, thereby reducing greenhouse gas emissions during the growing period of crops and reducing irrigation water, and improve production efficiency, thus contributing to the development of CSA in terms of formulating preventive measures and improving strategic decisions. The effect of the implementation of agrometeorological forecasts is generally good, but there is still a lot of room for improvement. Special training is also needed for the extension and intermediary agencies of agrometeorological services to enable them to play an active role better. Through effective training, farmers can become practitioners of climate intelligence, and provide solutions to farmers' problems, and finally give full play to the role of meteorological services. In order to better promote the role of agrometeorological services in the development of CSA, it is also necessary to upgrade agrometeorological services.

4.4.1.2 *Farmer Field School*

Farmer Field School (FFS) is a participatory extension method based on experience learning centered on farmers and their needs (FAO, 2002). It provides a low-risk environment for farmers to carry out experiments with new agricultural management practices, learn from their observations, acquire new practical knowledge and skills, and improve decision-making capabilities (Settle *et al.*, 2014). This method was originally developed by FAO and partners, aiming at the use of unsustainable pesticides promoted by many Southeast Asian countries in the context of the Green Revolution. It was first applied in Indonesia in 1989, which proved the potential of natural enemies to regulate pest populations in irrigated rice systems and introduce the concept of providing integrated pest management to farmers (FAO, 2002). Since then, Farmer Field Schools have been used in over 90 countries, initially in Southeast Asia, and later in sub-Saharan Africa, South America and the Caribbean, Near East and North Africa, Central Asia and Eastern Europe, and have been adapted to different crops, production systems and topics include sustainable agroecosystem management of vegetable crops, cereal crops and root crops (FAO, 2014), cotton-based systems (Settle *et al.*, 2014), integrated rice aquaculture systems (Geer *et al.*, 2006), livestock and agropastoral systems (Dalsgaard *et al.*, 2005; Okoth *et al.* 2013), tree crops, climate change adaptation, nutrition, linking to value chains, credit and savings or life skills, etc. In the past few years, Farmer Field Schools have also integrated elements of adaptation to climate change, such as FAO's Farmer Field School Integrated Plant and Pest Management Program (IPPM), which promoted the improved and adapted varieties and agroforestry in Mali and Niger Practices (FAO, 2015). Field schools in Indonesia have increased farmers' awareness of climate change and improved solutions to changes in rainfall patterns, such as recording and interpreting farm rainfall monitoring and field water harvesting (Winarto *et al.*, 2008). Farmer Field Schools usually have great potential for development because the content of Farmer Field Schools can easily adapt to farmers' needs in specific areas. However, in the context of CSA, another challenge involves the selection and decision of the Farmer Field School participants and other local stakeholders to determine and consider the best practices for climate change adaptation and mitigation in specific areas.

4.4.1.3 *Science Field Shop Solution*

The method of constructing agricultural extension services through interdisciplinary cooperation is committed to providing all participating farmers, extension agencies, and scientists with opportunities to exchange traditional and latest experience knowledge and scientific knowledge. This method includes regularly holding Science Field Shops (SFS, at least once a month in rural areas), where all parties participate in. This method has played a good role in the extension of agricultural meteorology. Agrometeorological learning refers to the learning of new weather and climate knowledge related to production by farmers. For scientists, new knowledge is not only farmers' traditional knowledge but also recent empirical knowledge. Farmers and local extension agencies are the targets of scientific entities. Over time, farmers' promoters are selected by farmers and undergo more in-depth training to play an important role in local agricultural upgrading. SFS conducted experiments in Yogyakarta, Indonesia, northern Java, and Lombok in 2008, 2010 and 2014, focusing on the extension of agrometeorological services, and achieved good results. The achievements could be concluded as follows: (1) helping farmers improve their understanding of what is happening and will happen in their agricultural environment; (2) improving the corresponding decision-making ability; (3) providing adequate training to intermediary agencies to better help farmers; (4) addressing the key challenge to maintain Science Field Shop Solution (SFS) season after season, with increasingly better-trained extension (and the withdrawal of scientists) to develop new problem-solving services. The SFS contents are dialogues between participants and the components discussed are new strategies, technologies and goals that provide new knowledge that can be used to tackle the vulnerabilities of local farmers and address related questions. These results showed that SFSs are suitable for CSA.

4.4.1.4 *Building Climate-Smart Farmers*

As people become more and more aware of value-added technical services that provide direct benefits to farmers and agricultural systems, the role of agricultural research and extension has become more and more important; global changes have made relevant agricultural technology and service extensions face more severe challenges. Cultivating climate-smart farmers is an important way to promote the sustainable development of CSA. In the "Twelfth 5-year Plan" of the Government of India, the government emphasized the importance of agricultural resilience and coordinated governance. Its main goal is to design and implement an all-inclusive participation strategy to reduce the vulnerability of agricultural communities and maintain their development. Community-level institutions and public leadership are key to the success of this initiative. The target groups of extension services in India are agricultural communities, especially poor farmers and women farmers, as well as elected and non-elected leaders of the local-level public governance system. Stakeholders include public administration mechanisms and decision makers, extension workers, members of civil society, cooperatives and farmers at the national, state, district, and village levels. The participation of a wide range of stakeholders is the key to the smooth implementation of the soil health plan and related climate-resilient agriculture initiatives. A new extension approach (KRISHI MAHOTSAV) was deployed in 2004 in India through the initiative known as the "Festival of Agriculture". This was a door-step approach, followed by special capacity-building programs on CSA, so as to provide guidance to farmers at the village level prior to the onset of monsoon. This approach introduces a soil health card to each farmer to facilitate crop selection and soil management based on soil health analysis. By screening the poorest farmers or animal breeders, the government provides appropriate certified seeds, pesticides, fertilizer mixes and sprayers, and the limit for each farmer is 1,500 rupees. Village communities have adopted a participatory approach, using silt dams and village ponds for water source protection, and encouraging farmers to use micro-irrigation. Maps are drawn from satellite imagery, and a team of agricultural scientists and an agricultural extension team visit each village, interact with farmers, and provide agricultural consultation before the monsoon breaks out. In fact, the Indian government implemented the "Soil Health Card" and comprehensive irrigation water management plan throughout India in 2015. However, it is necessary to strengthen contacts with elected and non-elected and public leadership to keep the plan going. The total income of the agricultural sector in Gujarat doubled between 2005 and 2013. For many years, despite the unstable monsoon and continuous drought, farmers' income and the growth rate have been maintained. Farmers choose target crops based on soil moisture content to reduce the use of chemical fertilizers and pesticides, and increase the rate of seed replacement. This indirectly helps in strengthening the management of the landscape, reducing the burden of greenhouse gases, further obtaining the support of water conservation, and reducing the energy consumption caused by water extraction. Due to the overlap of various information provided by different agencies to farmers, there are indeed gaps and problems in extension services. Taking India as an example, the challenges facing CSA include inadequate implementation of technology or alternative solutions in small farms, insufficient demonstration and sustainable delivery of services. The transfer of Indian extension services to CSA requires a quality multiplication model, which provides services to all farmers in relevant areas with the participation of the local agricultural extension administration.

4.4.1.5 *Climate-Smart-Type Risk Management*

In the context of more frequent extreme weather events and climate shocks, strengthening the early warning system provides an important guarantee for curbing the erosion of the development process of the rural sector. Allowing farmers to make farm management decisions through targeted climate information during the planting cycle will help reduce climate risks and avoid frequent food insecurity. Ninety percent of Senegal's agriculture rely on rainfall irrigation, but its rainfall varies greatly, especially in the northern regions, where crops are particularly vulnerable to unstable rainfall and long-term drought. Frequent extreme weather events in the context of climate change led to serious crop yield declines (Khouma *et al.*, 2013). With the support of the Consultative Group for International Agricultural Research (CGIAR) program on Climate Change, Agriculture and Food Security (CCAFS), about 3 million people across Senegal have received important seasonal precipitation and long-term weather forecast services, which are of great significance for making more informed agricultural management decisions. The climate information services provided enable farmers to improve their adaptability and increase agricultural productivity. The National Meteorological Administration (NMA) and the National Civil Aviation Administration (NCAA) of China have worked together to develop reduced-scale seasonal rainfall forecasts, improve the partners' ability to analyze long-term weather data and provide farmers with more feasible information. The widespread use and coverage of broadcasting, mobile phones and radio have made it possible to convey climate information to a large audience and promote the project in other parts of the country. Thanks to the partnership between CCAFS, the National Agency of Civil Aviation and Meteorology (ANACIM) ANACIM and the Union of Radio Association and Community of Senegal URACS, as well as the specific supporting and complementary roles of each stakeholder, the size of the Commonwealth of Independent States (CIS) has been expanded in 2014, the Ministry of Agriculture, Livestock and Food (MAGA) of Guatemala worked closely with the International Center for Tropical Agriculture (CIAT) and the CGIAR Research Program on Climate Change, Agriculture and Food Security (CCAFS) to develop the "CSA Prioritization Framework" (CSA-PF) and analyzed the investment opportunities of CSA in the Arid Corridor of Guatemala. The initiative seeks to identify CSA practices and priorities that will help strengthen the food security and livelihoods of vulnerable farmers in the region. In 2013, the Guatemalan government formulated an emergency plan to support drought-affected families in the drought corridor. The goal of CSA-PF is to build on this endeavor to develop a process that combines rapid response to climate risks with long-term planning to establish population adaptability affected by climate change and variability, and this framework has achieved good results.

4.4.1.6 *CSA Prioritization Framework*

CGIAR Research Program on Climate Change, Agriculture and Food Security (CCAFS) and the CIAT have designed an evidence-based CSA-PF. The CSA-PF is a participatory process that links multiple analytical tools and methods for evaluating CSA practices, including the following: methods for identifying climate-vulnerable agricultural areas and production systems based on adaptation, productivity, and mitigation. This component of CSA is led by experts to conduct quantitative and qualitative evaluations of the specific environmental results of CSA practices and technologies, conduct cost–benefit analysis of CSA practices, and evaluate the opportunities and challenges of adopting CSA. The main goal of CSA-PF is to provide decision makers at all levels (national and local governments, donors, non-governmental actors, and the private sector) with tools to help them plan the process and promote more accurate decision-making. Priority actions for CSA established at different business levels (national, regional, local) form a key output of the framework. To achieve this goal, CSA-PF constitutes a multi-stage participatory process that combines expert evaluation with feedback from national and local actors to ensure that priorities are aligned with needs. These stages are additional to guide and improve practice by screening a range of applicable CSA options. In the initial stage, CSA-PF users determined the scope of research, developed a large number of potential CSA options on this basis, and then used CSA-related indicators for evaluation. In the second step, stakeholders weigh the different CSA options to determine a shortlist of practices for further investigation. Then in the third stage, the economic costs and benefits of these practices are thoroughly evaluated. Stakeholders discuss the results together in the final stage and prioritize CSA practices for the portfolio.

CSA-PF has been used in the planning process from local to national levels. Pilot studies have been completed in Colombia, Guatemala and Mali, and continue to be used in Ethiopia, Ghana, Nicaragua and Vietnam. The MAGA in Guatemala is the first to use this framework to adapt CSA to the environment and stakeholders, such as government, academia, research centers, and production departments. The Guatemala pilot focuses on priority actions, and the next steps will include implementation, monitoring and valuation, and adjustments to the national plans.

CSA-PF aims to determine the priorities of target investments to achieve CSA's current and future goals. This approach allows the integration of sustainable development concepts into agricultural and climate change planning by expanding potential entry points for action and cooperation among actors working to achieve any sustainable development goals (productivity, adaptation, mitigation). The framework can also be tailored to adaptation or mitigation options in the agricultural sector. Facts have proved that CSA-PF helps to develop a vision that is jointly formulated by actors in the public, private and non-profit sectors. The vision is geared toward medium- and long-term agricultural development plans and takes into account multiple investment levels. However, the key factor for the success of this method is the commitment of stakeholders to continue participating in the process and their willingness to embark on the implementation of the portfolio on the ground.

4.4.2 *Technology Extension and Innovation*

4.4.2.1 *Digital Agriculture to Achieve the Goal of CSA*

Data have always been the basis of scientific research, and experimental design and statistical analysis have made great progress in the research of planting systems. Nowadays, the source, quantity, nature, purpose, and users of data are developing rapidly. The big data revolution has brought many opportunities and challenges in many areas of society (Mayer-Schonberger and Cukier, 2013). Farmers in developed countries usually use the Global Positioning System (GPS) sensors embedded in combine harvesters to draw yield maps with very high resolution. At the same time, cloud computing and automatic weather networks can collect climate data and provide these data in real time through a network platform. In the future, the advancement of unmanned aircraft, remotely connected sensors, satellite products, and the popularization of the internet will make agricultural production generate more data. This information can provide farmers and agronomists with highly relevant location-specific information to enrich their knowledge base and support more accurate decision-making.

Collecting farmers' data that describes crop management, yield and crop conditions and combining it with meteorological records and field soil data could help in finely describing the actual conditions of crop growth and the yield achieved. Using empirical modeling technology to mine the database, we can obtain the relevant crop's changes in environmental conditions, the main constraints and the best management practices in each situation. Clustering, principal component analysis, regression, and machine learning methods, such as artificial neural networks, classification and regression trees, are part of a portfolio of technologies that can be used to overcome the additional challenges (noise, sparseness, temporal changes and all factors). This process can be described as a large-scale benchmark test in which the performance of the crop is compared in several groups of farmlands with similar environmental conditions. Data-driven agriculture brings several opportunities: (1) quantifying crop yield gaps and their reduction pathways (van Ittersum *et al.*, 2012); (2) understanding in detail the response of different crop varieties to different meteorological factors at specific site scales; (3) categorizing the disaster events faced by crop production according to the characteristics of climate change; (4) realizing the best management practices for detection under specific environmental conditions.

Based on the results of the analysis, farmers can adjust their operations and management, and optimize input levels, variety selection, and sowing dates. Describing favorable or unfavorable climate patterns and their probability of occurrence is meaningful for them to predict crop rotation and manage risks. This method is particularly useful for small farmers who do not have access to extension services. This is a new tool for monitoring the large-scale response of crops to climate, soil and management. This system basically replicates the ideas of agronomists, but has accurate and reliable memory for computers. They will use it to enrich their suggestions based on actual measurement data from many farms. Plant breeders can use the results of the analysis to directly feedback on the performance of different varieties under commercial conditions. This will help them not only to adjust their breeding strategies based on the actual performance of the variety but also to design more and more local-specific varieties that are more adaptable to various environments. Agricultural organizations can obtain a lot of information about crop growth in order to optimize harvesting, logistics and commercialization.

Data-driven agriculture has been verified and implemented in some parts of the world, including China (Tao *et al.*, 2016), Colombia (Jiménez *et al.*, 2016), France (Delmotte *et al.*, 2011), Iran (Shekoofa *et al.*, 2011, 2014), as well in other countries and regions. Some countries, such as Argentina, Colombia, Mexico, Nicaragua, Uruguay and the United States, have also reported specific implementations under different business models. Some private companies have begun to develop solutions and services in these countries. Farmers provide their data through internet-based technology and receive personalized reports on the condition of their crops and suggestions on actions to be taken to make the most of their crops. For example, a climate company in the United States uses publicly available daily weather data to generate situational weather forecasts and management recommendations for member farmers. The Farmer Business Network also provides similar services, but it invites farmers to join a community in which farmers share their data anonymously and builds a huge knowledge database from which each member can learn and interact with millions of other people in the plan to perform benchmark management. Today, every John Deere machine includes a Subscriber Identity Module (SIM) card used to transmit operating data to the company in real time. These data include all information about the machine's running time, the frequency of failures, engine performance and many other factors. The company collects and analyzes these data to provide customers with reports on machine status, profitability, and suggestions on how to improve fleet management. In addition, the company also enters the field of agronomic consulting based on farmers' data. In addition, other research institutions are also carrying out related research work. For example, CGIAR centers are developing several initiatives to harness the benefits of farmers' data and trying to democratize its use. Among them are the Agricultura Especifca por Sitio (AEPS) team at CIAT and the Mas Agro project at the International Maize and Wheat Improvement Center (CIMMYT). A cross-cutting platform dedicated to big data is also planned to start in 2017. The Rothamsted Institute in the UK has invested heavily to make better use of big data in the agricultural sector.

Data-driven agriculture may take time to reach the critical amount of data that allows analysis. With the growth of data collection, there have been similar methods that rely on simpler information but use the same conceptual approach: Cropcheck in Chile, the Consorcio Regional de Experimentación Agricola (CREA) in Argentina or the Comprehensive and Economic Trade Agreement (CETA) in France (Gerbaux and Muller, 1984). The basis of these tools is to build collective knowledge through the benchmarks of performance member farms.

4.4.2.2 *New Type of CSA Extension Method: Plantwise*

Plantwise is a multi-partner global program for plant health system development launched in more than 30 developing countries in Africa, Asia,

Central America and South America, providing a new platform mechanism for coping with the many challenges resulting from climate change (Romney *et al.*, 2012). This is also a case that the African Soil Health Alliance is implementing the extended system, which can contribute to CSA in many ways. Plantwise aims to solve this problem by linking the mass media with the plant clinic and better communicate the information. Plant clinics work in a way similar to human health clinics; they are the frontline point of contact for the national extension system, allowing extension workers and farmers to directly exchange information on "any problem and any crop". Plant clinics are channels to promote face-to-face communication and two-way knowledge and information flow between extension workers and farmers, and to establish connections with other components of the plant health system (Boa *et al.*, 2015). They meet the urgent needs of farmers, provide advice as needed, and are regularly carried out by the state and local agencies in public places most suitable for meeting farmers' needs. So far, more than 1,600 plant clinics have been established in 33 developing countries in Asia, Africa and the Americas, where farmers can provide assistance on any plant health problem affecting the crops they grow and can get practical advice from extension workers. At the same time, the extension staff also filled out a so-called prescription sheet to record information about plant health issues and advice given. The access data of each farmer is stored in a central repository, which is a gold mine of real-time information. Plant clinics can directly or indirectly contribute to all three pillars of CSA. There is no doubt that when targeted recommendations are conducive to increased yields, plant clinics contribute to food security. Evidence from a recent study showed that 82% of farmers who visited plant clinics in Pakistan, Sri Lanka, Rwanda, Malawi, and Ghana reported that plant clinics increased crop yields (Williams, 2015). The increase in crop yields due to better crop management practices also means that better utilization of existing land and agricultural production resources improved production efficiency, helping to mitigate climate change. The role of plant clinics in adapting to climate change is mainly reflected in two aspects. On the one hand, due to climate change, the emerging speed of new problems is getting increasingly faster. Plant clinics connect with other plant health system stakeholders to provide a mechanism for quickly responding to new problems brought by farmers (Romney *et al.*, 2013). On the other hand, plant clinics can play a key role in monitoring. Since the model cannot accurately predict the impact of climate change, it is difficult to formulate and launch climate change adaptation measures for specific locations. With the data collected in the plant clinic system, unexpected crop production problems caused by climate change can be discovered. This allows the government to formulate appropriate response strategies, such as through plant clinics or other extensional methods, to better respond to emerging pests, temperature increases or changes in the growing season. Plant clinics can also play an important role in establishing the resilience of the entire agricultural system, as their recommendations focus on integrated pest management principles. This approach reduces the dependence on agricultural inputs (fertilizers and pesticides), so that smallholder farmers can achieve economic recovery faster after the price of these inputs rises.

The plant health rally method is an extension method that can quickly raise people's awareness of the major agricultural risks or threats facing important crops, promote the adoption of improved agricultural practices, and collect farmers' feedback on major issues affecting production. The plant health rally method is a supplement to the plant clinic method, and the information they transmit varies in scope, impact, and complexity. They are usually held in public places and are open to everyone. Plant health rallies may be spontaneous, attracting people with flags and other announcements, or they may be aimed at farmers who are specifically mobilized for this event. Each rally starts with a brief explanation of the chosen topic, and then people (mainly farmers) can ask questions and get a fact sheet of verified recommendations. The different synergy between the plant clinic and the plant health rally can be determined. Most obviously, the data collected in the plant clinic can be used to determine the theme of the plant health rally, making this method an extremely sensitive and powerful extensional tool. By the end of 2014, nearly 290 plant health rallies had been held in 14 different countries, and targeted messages were delivered to more than 21,000 farmers. Especially in countries such as Malawi, Uganda, and Zambia, this method has been adopted by public extension systems and has been valued because it can reach a large population in the target area in a short period of time (Mur *et al.*, 2015). Experience from Malawi shows that 34 plant health rallies were held in two days, and more than 4,000 farmers have reached a management strategy on important pests and diseases such as cassava mosaic virus or witchweed that pose a huge threat to food security.

The goals of the plant health rally and the plant clinic are the same. The focus of the plant clinic is mainly on emerging problems and making targeted recommendations to promote food security and alleviate climate impacts, but the plant health rally places more emphasis on raising awareness and prevention. The plant health rally method does not mean face-to-face communication between individual farmers and extension workers. For example, in plant clinics, the complexity of the information delivered to farmers is reduced, but the number of farmers reached increases. Due to its mixed nature between mainly supply-led methods (such as classic mass media methods) and two-way demand-driven methods (such as plant clinics), people have a better understanding of climate change and its related effects on agriculture. This approach also helps to mitigate climate change, for example, the plant health rally will find a perfect tool to mitigate climate impacts. An example is the fact that the use of urea deep burial technology in rice production can reduce greenhouse gas emissions from rice.

4.4.2.3 *Public Science Approach for CSA: Triadic Comparisons of Technologies*

Triadic comparison of technologies (TCICOT) is an approach that applies citizen science and crowdsourcing methods to evaluate climate-smart technologies on-farm in a way that facilitates joint learning across different sites. The TCICOT citizen science method is participatory in nature and can be extended to many farmers. This method uses a simple technical evaluation method to evaluate many farmers. Each farmer accepts the three technologies of seeds and fertilizers and feeds back from the farmers' perspective on which technology is the best and the worst among the three technologies. Since each farmer receives a different combination of the three technologies, it is possible to evaluate a combination of multiple options. By combining data generated by farmers with data from other sources, especially from climate, soil and socio-economic conditions, this method can obtain information about the interaction of technology with its environment and its potential contribution to CSA. This information can help introduce farmers to various options through field activities and provide information to researchers, inputs and service providers. The method is designed on the basis of extensive experiments conducted in India, East Africa and Central America. The experimental project is led by the International University of Biology. Its purpose is to promote the Global Climate Change Institute (GCCI) to strengthen research programs on climate change, agriculture and food security. Many organizations have now implemented this method, including the Centro Agronómico Tropical de Investigacion Enseanza (CATIE) (Central America), the

National Association of the Remodelling Industry (NARI) (Tanzania), Mekelle University (Ethiopia), ICAR and a large number of Krishi Vigyan Kendra KVK farm science centers, which are tasked with conducting farm tests (India). Some other organizations are also preparing or considering conducting this experiment.

4.4.2.4 *The Role of Early Warning Systems in Promoting CSA*

The early warning system helps community farmers improve their resilience to weather patterns and food insecurity, and provides farmers with accurate forecast information in a timely manner. The Kotido region in Uganda, Africa, is in a state of drought, insecurity and food insecurity all year round, which significantly restricts human health and sustainable agricultural development, leading to many problems such as water supply and frequent occurrence of crop pests and diseases. Local and national governments, UN agencies, the Famine Early Warning Systems Network (FEWSNET) and other agencies have worked with development partners to design an early warning system for the entire Karamoya region. The system is implemented under the existing structures of local and national governments, and a non-governmental organization called ACT works with the Prime Minister's Office and the Ministry of Agriculture to provide technical and financial support for its operations. The system includes analyzing and reviewing the data collected every month, and disseminating it to at-risk communities in the region, district and sub-counties. By monitoring selected indicators, risks such as drought, famine, outbreaks of pests, sales, and unsafe drinking water for humans and animals can be predicted in advance. Therefore, whenever the disaster risk rises, the system can promptly send warning signals to communities, relevant regional departments and development partners, and implement disaster preparedness measures to minimize the impact on the population. The heads of various departments in the district can make suggestions to various stakeholders and clarify the best strategies to help the communities get prepared.

4.4.3 *Knowledge and Sharing*

4.4.3.1 *Digital Green*

Digital Green started in 2006 as a research project of the Emerging Markets Technology Group of Microsoft Research (EMTGMR), whose purpose is to understand how information and communication technologies support small agricultural systems. Digital Green tested multiple methods and compared the use of posters, audio messages, and videos with traditional extension methods, such as training and visits (extension staff visit farmers and personally provide training to improve practices) and farmer field schools (using demonstration fields to provide practical training and demonstrations of improved practices). These experiments have promoted the development of Digital Green methods that cooperate with existing public, private, and civil society organizations to participate in the extension and layering of technologies to enhance their effectiveness and efficiency. In 2008, Digital Green became an independent non-profit organization, and through adjustments and strengthening of enlarged promotion and application, more than 800,000 farmers (80% of women) in India, Ethiopia, Afghanistan, Ghana, Niger, Papua New Guinea and Tanzania were involved.

Digital Green adopts a participatory method of extension agency training. It uses low-cost pocket cameras, microphones and tripods to share recommendations and short videos made by fellow farmers, allowing local farmers to demonstrate improved agricultural practices. By watching videos and discussing, it records farmers' feedback, their questions and interests, and data on the practices they adopt. Data and feedback help improve the production and distribution of the next set of videos in the iteration cycle to further meet the needs and interests of the community. In a controlled evaluation, it was found that the efficiency of the new method was 7 times that of the original, and the efficiency was 10 times that of the original at the same cost (Gandhi *et al.*, 2009). Participatory video is the core of the Digital Green approach; Digital Green can also use other communication channels, such as radio, television, and mobile applications to disseminate information and connect farmers to the market. Digital Green discovers that different communication methods can complement each other on another level. This is the productivity continuum of cognitive knowledge adoption–agricultural extension. When the information comes from sources similar to their own, farmers' acceptance is higher. Digital Green also uses videos to build courses and incorporates them into the training system of extension agencies. Digital Green's model promotes the adoption of CSA practices in India and Africa, which can sustainably increase agricultural productivity and income. Most of the video content produced and distributed to farmers focus on farm management and agronomic practices that increase farm productivity, rather than traditionally promoted "technology transfer", which often focuses on providing improved inputs, including harmful synthetic pesticides and fertilizers. It also reduces farmers' costs, improves farmers' resilience, and reduces their risk of climate and market shocks. Emphasis is placed on the use of local existing and endogenous knowledge and resources, as well as those that have the greatest limit on production, including water, soil management and pest management. For example, the National Rural Livelihood Mission (NRLM) of India is using Digital Green methods to promote improvements in rice production practices, including nursery cultivation and transplantation in rice planting, weeding and water management, and seed treatment. These practices follow the environmental guidelines approved by the government, which include pest management, disease management, soil nutrition management, planting patterns in rain-fed areas, and soil and water conservation. Following these guidelines is an important part of Digital Green's approach to working with existing systems, as the establishment of parallel systems limits the long-term viability of this approach.

4.4.3.2 *Climate Information Services*

The training of agricultural meteorological institutions is a crucial link in the development of CSA. The extension agencies generate useful knowledge for agricultural production decision makers through training, and most of them were specially trained staff from the NMA and extension departments of universities and research institutions. These departments provide content for responding to agricultural scientific support systems (Stigter *et al.*, 2013), such as weather and climate forecasts, drought, flood or other disaster monitoring and early warning products, and agricultural meteorological service consulting. These services can improve farmers' preparedness, but they must be made into products that are convenient for customers to use in rural areas to respond to climate change (Stigter *et al.*, 2013). This kind of response in agriculture must be developed and solutions to farmers' problems should be sought at the level of applied science. In terms of training, this type of "product intermediary agency" needs to have a good education on the needs of farmers and how to use the knowledge of applied agrometeorology to promote and solve problems in agrometeorology. Another type of intermediary agency is the closest to farmers. They should learn to better express the needs of the farmer community and look for agrometeorological components that need attention. They should match this with existing or forthcoming agrometeorological services and maintain close contact with product intermediaries. The above examples of agrometeorology can also be applied to agricultural hydrology, agricultural entomology, agricultural pest management, crop management, farm management and other fields after proper modification.

4.4.3.3 *Farmer Organizations and CSA: Vietnam Case*

Vietnam is a large agricultural country. Two-thirds of the population live in rural areas. On the basis of ensuring national food security and forest coverage, Vietnam also outputs a lot of rice, coffee, rubber, cashew nuts, black pepper, wood and aquatic products, and achieves remarkable results in agricultural production. Vietnam's agriculture faces many problems, like low profit margins, high labor input caused by obsolete equipment, small and scattered farmland, poor infrastructure and investment options, and weather-related pressures. Vietnam has two major systems: (1) Public extension is conducted through the Ministry of Agriculture and Rural Development (MARD). In remote areas, there is usually only one extension officer in charge of the entire commune. (2) The Farmers Union (FU) is a civil society organization under the Party and the People's Committee. Among these farmers and promoters, an increasingly popular activity is private extension activities linked to the sale of inputs, such as seeds and fertilizers. MARD has participated in CSA since 2010 and hosted the Second Global Conference on Agriculture, Food Security and Climate Change in Hanoi in 2012. MARD believes that participation in GACSA will add value and sustainability to agriculture. For example, FAO's Economic and Policy Innovations for CSA Program works with MARD on the CSA value chain of indigenous products. Agriculture and forestry have strong characteristics in the expected contribution to the UNFCCC determined by the country. CSA initiatives include rice-sustainable intensification and alternating wetting and drying in delta regions (Siopongco *et al.*, 2013) and forestry through payment of forest environmental services (PFE). In 2014, through the CGIAR Research Program on Climate Change, Agriculture and Food Security (CCAFS), the Climate-Smart Village (CSV) was introduced. Six villages were selected to represent different climate-exposed agricultural ecosystems, including three in Vietnam, two in Laos, and one in Cambodia. CSV is a test site for scalable CSA technology, based on the partnership between CGIAR and local institutions. The development of CSA in Vietnam faces three challenges: (1) CSA has different interpretations; (2) the support policies are not clear; (3) the participation time of women in training cannot be guaranteed. The main opportunities for developing CSA in Vietnam are as follows: (1) Export crops are mainly grown as monoculture or short-term rotation, but they can all be grown in an integrated system. (2) It is possible to work with extension staff and farmers to develop "smart" indicators to monitor adaptation, reduce or stabilize production, and increase income. These indicators can also monitor extended presentations. Studies have shown that agroforestry management shortens the economic recovery period after natural disasters. (3) With a 3–4-year rotation program, it is vital to cooperate with those who stay and those who learn to create resources. (4) MARD is essential to policies that promote expansion.

References

Boa E, Danielsen S, Haesen S. 2015. *In One Health: The Theory and Practice of Integrated Health Approaches*, pp. 258–259.

Boa E, Danielsen S and Haesen S. 2015. 'Better together: identifying the benefits of a closer integration between plant health, agriculture and One Health.', CABI Books. CABI. pp. 258–259.

Carter T R, Jones R N, Lu X, *et al.* 2007. New assessment methods and the characterisation of future conditions. *Climate change 2007: Impacts, Adaptation and Vulnerability*, pp. 133–171.

Colorado State University. 2013. Agriculture and land use national greenhouse gas inventory software.

Dalsgaard J P T, Minh T T, Giang V N, *et al.* 2005. Introducing a farmers' livestock school training approach into the national extension system in Vietnam.

Davis K. 2009. *Agriculture and Climate Change: An Agenda for Negotiation in Copenhagen.*

Delmotte S, Tittonell P, Mouret J-C, *et al.* 2011. On farm assessment of rice yield variability and productivity gaps between organic and conventional cropping systems under Mediterranean climate. *European Journal of Agronomy*, 35 (4): 223–236.

Dessai S, Hulme M. 2004. Does climate adaptation policy need probabilities. *Climate Policy*, 4 (2): 107–128.

Engle N L. 2011. Adaptive capacity and its assessment. *Global Environmental Change*, 21: 647–656.

FAO. 1988. Definition of several extension approaches.

FAO. 2002. From farmer field school to community IPM — Ten years of IPM training in Asia.

FAO. 2010. Greenhouse gas emissions from the dairy sector: A life cycle assessment.

FAO. 2012a. Using marginal abatement cost curves to realize the economic appraisal of CSA: Policy options.

FAO. 2012b. Incorporating climate change considerations into agricultural investment programmes: A guidance document.

FAO. 2012c. Learning module 2: FAO approaches to capacity development in programmes: Processes and tools.

FAO. 2014. Transboundary Agro-Ecosystems Management Project for the Kagera River Basin.

FAO. 2015. Building resilient agricultural systems through farmer field schools — Integrated Production and Pest Management Programme (IPPM).

Folke C. 2006. Resilience: The emergence of a perspective for social-ecological systems analyses. *Global Environmental Change*, 16 (3): 253–267.

Gandhi R, Veeraraghavan R, Toyama K, *et al.* 2009. Digital Green: Participatory video and mediated instruction for agricultural extension. *Information Technologies and International Development*, 5 (1): 1–15.

Geer T, Debipersaud R, Ramlall H, *et al.* 2006. Introduction of aquaculture and other integrated production management practices to rice farmers in Guyana and Suriname.

Gerbaux F, Muller P. 1984. La naissance du développement agricole en France. *Economy Rural*, 159: 17–22.

German International Development Cooperation (GIZ). 2011. Making adaptation count: Concepts and options for monitoring and evaluation of climate change adaptation.

GFRAS. 2012. The "New Extensionist": Roles, strategies, and capacities to strengthen extension and advisory services.

IPCC. 2012. Managing the risks of extreme events and disasters to advance climate change adaptation. A Special Report of Working Groups I and II of the IPCC, 582 pp.

Jiménez D, Dorado H, Cock J, *et al.* 2016. From observation to information: Data-driven understanding of on farm yield variation. *PLoS One*, 11 (3): e0150015.

Khouma M, Jalloh A, Thomas TS., *et al.* 2013. Senegal, pp. 291–322.

Mayer-Schonberger V, Cukier K. 2013. Big data: A revolution that will transform how we live, work and think, 242 pp.

Mloza Banda BC. 2014. E-adaptation to Climate Change in Malawi, ICT Update 78.

Mur R, Williams F, Danielsen S, *et al.* eds. 2015. Listening to the silent patient. Uganda's Journey towards Institutionalizing Inclusive Plant Health Services, 224 pp.

Nelson G. 2009. Agriculture and Climate Change: An Agenda for Negotiation in Copenhagen. *International Food Policy Research Institute* (IFPRI), 2020 vision focus.

Okoth JR, Nalyongo W, Petri M, *et al.* 2013. Supporting communities in building resilience through agro pastoral field schools.

Romney D, Boa E, Day R, *et al.* 2012. Solving the problem of how to solve problems: Planning in a climate of change.

Romney D, Day R, Faheem M, *et al.* 2013. Plantwise: Putting innovation systems principles into practice. *Agriculture for Development*, 18: 27–31.

Settle W, Soumaré M, Sarr M, *et al.* 2014. Reducing pesticide risks to farming communities: Cotton farmer field schools in Mali. *Philosophical Transactions of the Royal Society of London*, 369: 20120277.

Shekoofa A, Emam Y, Shekoufa N, *et al.* 2014. Determining the most important physiological and agronomic traits contributing to maize grain yield through machine learning algorithms: A new avenue in intelligent agriculture. *PLoS One*, 9 (5): 97288.

Siopongco JDLC, Wassmann R, Sander BO. 2013. Alternate wetting and drying in Philippine rice production: Feasibility study for a Clean Development Mechanism.

Sones KR, Oduor GI, Watiti JW, *et al.* 2015. Communicating with small-holder farming families: A review with a focus on agro-dealers and youth as intermediaries in sub-Saharan Africa. *CAB Reviews*, 10 (030): 1–6.

Stigter CKJ, Winarto YT. 2013. Science field shops in Indonesia. A start of improved agricultural extension that fits a rural response to climate change. *Journal of Agricultural Science & Applications*, 2 (2): 112–123.

Tao F, Zhang Z, Zhang S, *et al.* 2016. Historical data provide new insights into response and adaptation of maize production systems to climate change/variability in China. *Field Crops Research*, 185: 1–11.

UNFCCC. 2010. Handbook on vulnerability and adaptation assessment. Consultative group of experts on national communications from parties not included in Annex I to the convention.

United Nations Environment Programme (UNEP). 2013. Carbon benefits project: Modelling, measurement and monitoring.

Van Ittersum M K, Cassman K G, Grassini P, *et al.* 2012. Yield gap analysis with local to global relevance — A review. *Field Crops Research*, 143: 4–17.

Villanueva P S. 2010. Learning to ADAPT: Monitoring and evaluation approaches in climate change adaptation and disaster risk reduction — challenges, gaps and ways forward.

Walker B, Gunderson L, Kinzig A, *et al.* 2006. A handful of heuristics and some propositions for understanding resilience in social-ecological systems. *Ecology and Society*, 11 (1): 13. http://www.ecologyandsociety.org/vol11/iss1/art13/

Walker B, Holling C S, Carpenter S R, *et al.* 2004. Resilience, adaptability and transformability in social-ecological systems. *Ecology and Society*, 9 (2): 5. http://www.ecologyandsociety.org/vol9/iss2/art5/

Williams F. 2015. Plantwise Rapid Satisfaction Survey.

Winarto Y T, Stitger K, Anantasari E, *et al.* 2008. Climate Field Schools in Indonesia: Improving "response farming" to climate change.

World Bank. 2009. Guidance notes mainstreaming adaptation to climate change in agriculture and natural resources management projects.

Chapter 5

International Climate-Smart Agriculture Cases

Abstract

This chapter compares and analyzes typical cases of climate-smart agriculture (CSA) in developed and developing countries and summarizes the existing problems and challenges of CSA. The case study in Italy, representing developed countries, mainly introduces the impact of climate change on Italian agriculture and economy, the effect of forecast on CSA there, and the opportunities and challenges that Italy is faced with in the development of CSA. In Mozambique, a representative of the developing countries, the development of CSA is much poor than that of Italy. For Mozambique, this chapter gives the introduction on the impact of climate change on its agriculture and food security, mainly applied CSA technologies, practices, systems, and policies. Based on the introduction and study of these two cases, this chapter emphasizes the characteristics and development priorities of the climate-smart crop production system, which includes crop variety improvement, construction of diversified planting systems, climate-smart soil management, climate-smart irrigation, integrated pest and disease management, agricultural mechanization level, and weather forecast. The mature CSA production systems such as European crop rotation system, American soybean-maize system, and rice production system are briefly introduced. The objective of this chapter is to provide relevant support for China's CSA crop production system.

5.1 Case Study of CSA in Developed Countries: Italy

5.1.1 *The Impact of Climate Change on Italian Agriculture and Economy*

Italy is an important agricultural producer in Europe. The climate and soil conditions in Italy vary in different areas significantly. Moreover, Italy faces many environmental challenges such as soil loss, desertification, erosion, general degradation of the ecosystem, and escalation of extreme events. Italy has nearly half of Europe's plant species and more than 58,000 animal species, which makes it almost having the highest biodiversity density in Europe. In the past few decades, due to the development and support of sustainable policies, although the Italian national agricultural sector has undergone huge changes in response to climate change and agriculture, the dialogue between and participation of various sectors are developing constantly. The promotion of cohesive policies, technical panels, and capitalization of experiences are still needed to keep these ongoing. Statistics from the agricultural sector show that agriculture provides 429,000 jobs with an added value of 3.3% in agriculture, forestry, and fishery sectors, and the output value is estimated at €54,438 billion (Istat, 2015). Although the total agricultural area is just about 12.86 million hectares, the agricultural area accounts for 2.4% of soil erosion. The area of soil erosion has increased from an average of 7.9–8.4 hectares per year. It is relevant that the number of multifunctional farms operating in agriculture has recently increased (+48.4%) due to the rise of farms producing renewable energy (21,000 farms) and converting their products (+97.8%). However, the livestock sector is in decline for pigs (−7.8%), cattle (−4.5%), and poultry (−1.5%), while sheep farms increased slightly (+0.5%). Italy has developed organic agriculture, with more than 9% of agricultural land used for organic production, ranking fourth in Europe and sixth in the world. The output of certified organic products has increased significantly (+7.7%) and use of pesticides has been declining (FAO, 2016a).

The national rural policies have been developed starting from 1996 with the adoption of the European Rural Chart (EDS), which links the use of agricultural land with environmental and socio-economic aspects. The European Union applied some directives and regulations, such as the Sustainable Use of Plant Protection Products (Directive 2009/128/CE) and the European Development Strategy (EDS), which have further strengthened the regional and national implementation of sustainable actions in the agricultural sector. The reform of the Common Agricultural Policy (CAP) began in 2013, which has not only expanded the scale of Italian commercial farms in the past 10 years, but large companies have also become more efficient and competitive in adapting to changes in production and demand. The Italian socio-economic framework is still characterized by a structural weakness as a result of the current general global financial crisis, resulting in a low growth performance and an economic contraction estimated at 0.9% of gross domestic product (GDP) annually (Istat, 2016), although it is remarkable that the agricultural sector supports Italian GDP with 1.8% of added value. As a result, the decline in agricultural income is still significant. Compared with other European Union (EU) countries, Italy's agricultural income has decreased by about 6% in the past 10 years (Pasqui *et al.*, 2012). But Italy still leads the way in the output of high-quality agricultural products.

Italy actively participated in the implementation of policies related to climate change, approved the Paris Agreement (COP21) and national law 2014/2016. The Italian Ministry for Environment, Land and Seas (IMELS) approved the national strategic document "Strategia Nazionale di Adattamento ai Cambiamenti Climatici" (SNAC) for adaptation to climate change in 2015. The SNAC mentioned that agriculture and food production largely depend on the condition of natural resources, and they are particularly sensitive to the effects of climate change. Knowledge of the status, impacts, and vulnerabilities of climate change is essential for the local agricultural sector to define and implement the most appropriate adaptation measures. Many agricultural areas in Italy are vulnerable to climate change. The introduction of appropriate adaptation strategies intends to minimize their vulnerability and help increase food production and mitigate climate change. Within the national strategy, SNAC, a forum to promote information, knowledge, and citizen's participation, has been established in addition to a permanent observatory "Osservatorio Nazionale" composed of representative of regional and local institutions and stakeholders, in order to identify priorities, support decision-making, and monitor the effectiveness of actions and policies. The "National Plan for Adaptation to Climate Change" will identify priority actions in key areas of SNAC. In this regard, several best practices can be provided in different plans through projects funded by the European Union (EU), such as CIP Eco Innovation and CIP Intelligent Energy Europe (IEE). The knowledge platform website (Piattaforma delle Conoscenze) shares these best practices and facilitates access to them, stimulating networking activities between relevant actors.

5.1.2 *Italian CSA*

5.1.2.1 *CSA at the National Level*

Italy pays great attention to the formulation and implementation of CSA-related policies. Relevant policies and programs that promote the development of CSA have played an important role in practical application in Italy, and related programs have been incorporated into relevant European directives and regulations. Within the national framework, the main policies and financial instruments that contribute to mitigating climate change mainly come from CAP, especially from the pillar dedicated to rural development measures. Agriculture is directly related to these emission mitigating measures, accounting for 24% of greenhouse gas emissions and also acting as a potential reservoir for carbon storage. The second sponsor of CAP is the European Agricultural Fund for Rural Development (EAFRD). In Italy, the plan is implemented by 21 Regional Rural Development Programs (RRDP) to help farmers take agricultural management measures that are beneficial to climate and environment. The major national entities enabling and supporting CSA policies with a distinct portfolio along the territory are the Ministry of Agriculture, Food and Forestry Policies (MiPAAF), and the Italian Ministry for the Environment, Land and Sea (IMELS), national research institutions such as the Italian National Institute for Environmental Protection and Research (ISPRA), the National Research Council of Italy (CNR), the Council for Agricultural Research and Agricultural Economics Analysis (CREA), and some autonomous provinces at the local level are also involved. It should also be noted that MiPAAF is managing a National Rural Network (NRN) plan implemented by CREA for the period 2014–2020, which aims to support national and regional authorities in the implementation of EAFRD policies by incorporating RDP's CSA themes.

In Italy, the leading projects, programs, and funds have promoted CSA at different policy and technological levels, while creating effective synergy and complementarity among various actors and drivers. Specifically, through the formulation of policy and special management funds, research and innovation policies are implemented at both the national and regional levels. Nationally, MiPAAF, IMELS, the Ministry of Economy and Finance (MEF), and the Ministry of Education, University and Research (MIUR) are the main institutions dealing with agri-food research. Agricultural research is regulated by specific rules within a particular area. While the coordination of an interregional network of agricultural research, forestry, aquaculture, and fisheries plays an important role between areas. In the field of development services and innovation transfer, the regional administrations have full autonomy of action. The "Strategic Plan for Innovation and Research in Agriculture, Food and Forestry (SPIRAFF)" describes MiPAAF and Italian regions' strategy to meet the dictates of the first priorities envisaged by the EU Regulations for the 2014–2020 programming period of the CAP, in order to promote the transfer of knowledge and innovation in agriculture and forestry in rural areas, within the framework of the rural development policy scheme. The concrete achievements of the European Innovation Partnership (EIP) from the opportunities offered in different areas are embodied in research and innovation policies, national cohesion policies, environment and climate change policies, consumer and health policies, education and training policies, and industry and information policies.

The main objective of the agricultural sector is to respond to challenges and promote research, innovation, and training at the national and local levels. According to the European Plan (2014–2020), Italy's policies focus on thematic goals, with a special emphasis on strengthening research, technological development, and innovation; improving the competitiveness of small and medium-sized enterprises in the agricultural sector; developing fisheries and aquaculture sector; promoting the prevention and risk management of adaptation to climate change; protecting the environment and promoting the effective use of resources. At the national level, Italy has formulated a number of plans and projects, and joint actions have been taken to address sustainable development priorities. The MiPAAF focuses on conservation agriculture (CA), which plays an important role in CSA. And it also protects the promotion and application of CA in Italy through the special support measures envisaged by the regional RDP and strives to cultivate farmers who voluntarily switch to conservation farming practices. CA practices are a valuable mitigation strategy for the excessive consumption of soil and resources, while actively maintaining soil fertility and biodiversity. The promotion of CSA will help protect arable land, improve social sustainability and resource utilization, and optimize production in conjunction with external inputs. The main components of CSA in Italy include technical interventions such as direct seeding and no-tillage seeding, removal of plowing systems, cover crop system between two consecutive plantings, and leaving residues in the fields as mulch. Minimum tillage and strip tillage require the treatment of the top 15 cm of soil besides vertical tillage (5 cm and 8 cm). Through the special projects arranged by the NRN 2014–2020 plan, MiPAAF provides firm support to regions and stakeholders, aiming to facilitate the adoption and spread of CSA under the support of the EAFRD.

5.1.2.2 *Investment Environment for CSA in Italy*

European countries attach great importance to the support of public finances for sustainable development, especially in dealing with the new challenges caused by climate change. The Common Strategic Framework (CSF) aims to provide long-term socio-economic and environmental assistance from 2014 to 2020. European priorities have been coordinated with the needs of Italy and specific sectors, ensuring cooperation, convergence, and competitiveness. Research and innovation, as well as education and social cohesion, have been reinforced, shifting to a resource-efficient, low carbon economy.

The European Union EAFRD budget amounts to €99. 6 billion. This accounts for approximately 24% budget of CAP. In the period 2014 and 2020, the expected total public expenditure of the EU's rural development policy is €161 billion. In the EU's EADFR budget, the environmental and climate priority areas account for 52% of public funds, about €80 billion. Agri-environment-climate payments (AECP), established by Measure n. 10 of the RDPs, broadly represent 17% of the EAFRD budget in the EU. Italy's budget for 2014–2020 RDPs is approximately €21 billion roughly and 41% of their budgets are devoted to environmental and climate priorities. In particular, the sole Measure n. 10 accounts for a budget of 12% of the entire planned spending (€2.3 billion for 2014–2020, which is the planned quota estimated to be allocated in 7 years), with availability of funds aiming at incentivizing sustaining farmers and other land managers for the preservation of natural resources, genetic patrimony, the delivery of ecosystem services, and climate change mitigation or adaptation actions.

According to the analysis of the national rural network, the impact of public spending on RDP structural measures for innovation has considerably increased between 2007 and 2015 on the value of gross fixed investment in the agricultural sector, rising from 8% to 12%. In addition to the European

Programme Horizon 2020, which allocates more than €3.8 billion to "Food security, sustainable agriculture, marine and maritime research and bio economy", the major funds devoted to CSA derive from the CAP and especially from the EAFRD.

Private investment is an auxiliary tool of state funds. One example of this is the multi-regional guarantee platform of Italian agricultural sector companies providing a total of €465 billion in funding since 2017. This is the first experimental example of private state funding supported by MiPAAF and involving cooperation between regional authorities, national institutions, and European financial institutions. The multilateral investment portfolio aims to protect and fund RDP-related loans and investments, while using EAFRD's support to ensure that small and medium-sized enterprises (SMEs) are protected not only in the process of production, processing, and distribution, but also in the process of transition and application to CSA.

The cohesion of political funds has also generated many resources, such as the Development and Cohesion Fund (FSC), the National Operation Plan (PON), and other funds issued throughout the international calls, including ERA-NET and the European Union Joint Programming Initiative (JPI). According to Italian Law No. 499/199, other resources at the national level are related to funds used for intervention in the agricultural sector. Concerning the promotion of CSA practices at the regional level, 15 out of 21 Italy's RDPs have envisaged a number of support measures adapting them to pedoclimatic territorial characteristics. They aimed at helping farmers in the shift towards CA practices such as no tillage or minimum tillage, and other sustainable farming practices for promoting soil health. In fact, the transition from agronomy to no-tillage has increased the costs associated with new farming techniques using new agronomic management techniques. In Italy, between 2014 and 2020, the resources allocated for this typology of interventions account for €500 million, with a target intervention area of 330,000 hectares through the 7-year programming.

5.1.2.3 *Technologies, Practices, and Services Related to Italian CSA*

In recent years, Italy has implemented policies and services within the framework of "blue agriculture". These projects contribute to the common goal of responding to climate change vulnerabilities with their peculiarities exponentially, while stitching together national actors in a sort of

"Glue Agriculture". With the actions of the NRN program managed by MiPAAF, the Research Center for Agricultural Policies and Bio-economy of CREA (CREA-PB), in partnership with IMELS, regions and some national environmental organizations (NGOs), is particularly effective in the development of activities aimed at supporting CSA priorities within the RDP, as well as improving the efficient use of resources arising from EAFRD. In particular, the ongoing project NRN CREA 5. 1, "Actions in support of agro-climate environmental priorities of RDP (2016–2018)", focuses on CA within the scope of achieving sustainability and cost-effective benefits. The project also aims to improve the livelihood of farmers through the application of the three main CA principles, including minimal soil disturbance, permanent soil cover, and crop rotation. It also encourages to build networks and to enhance knowledge sharing on techniques and policies fostering rural development and stakeholder engagement.

The promotion of information and communication networking activities among various national stakeholders, supported by research and innovation actors, is fundamental in encouraging the sharing of best practices and knowledge not only on CA and CSA but also with regard to land and degradation neutrality. It is one of the crucial topics of sustainable development objectives to 2030. Moreover, at the national level, the project is particularly useful in the analysis of the complementarities and limits of rural development policies, not only within the Italian territory but also in other regional and European actions on crucial themes, such as land degradation, field type change, climate change adaptation, emission mitigation, biodiversity, and sustainability promotion. Through cooperation between FAO, OECD, government departments, and international organizations, as well as participatory activities, exchanges, and advocacy, the complement and coordination of common country assessment goals have been strengthened.

CA is an important practice of soil protection and practices such as no-tillage and minimum tillage are also envisaged as compulsory commitments to be adopted within the broad Integrated Production (IP) certification scheme. It is one of the major target policies in the RDP measure. In fact, there are approximately 70,000 hectares of land for integrated production, and funds have been allocated to maintain farmers' transition in planting herb and arbor crops and to adopt this protection technology. Since 2015, the European Institute of Technology (EIT) has been co-funding the Climate Smart Agriculture Booster (CSAb) flagship project with Climate Kic (Sustainable Land Use). CSAb's mission is to become

Europe's leading CSA innovation platform, knowledge portal, and market-place, incubating scientifically validated innovations, accelerating the adoption and the scaling of solutions, and facilitating transition to a climate-sustainable agricultural sector across Europe and beyond. The Biometeorology Institute of CNR, as the Italian partner of a restricted CSAb core group, is active at the country level in CSA boosting activities. The implementation of CSAb at the national level operates at different scales, spanning from the construction of an innovation platform to build and manage the users' community, and then extend to the definition of CSA solutions and services, as well as the development of a regional CSA centre for the engagement of national stakeholders and partners.

5.1.2.4 *Italy CSA Extension Services*

The key elements in capitalization and diffusion of innovations and best practices aiming at cementing sustainability are the awareness and knowledge of farmers, since the success of CA and other climate-smart practices depend on the appropriation from the whole actors and their active participation. Cultivating farmers for lifelong learning contributes to sustainable production technology. The average annual income of young Italian farmers is more than €50,000, accounting for 23% of farm manager incomes. They are very representative in Italy and have important economic contributions. Since 2010, a new trend has emerged in the Italian countryside, the return of young people who decided to engage in agriculture after university studies. If compared to previous generations, these "young farmers" possess expertise and knowledge (i.e. some of them are agronomists) and are more innovative in terms of new technologies, thus they have a high economic potential, which has been estimated at 40% more than their senior colleagues. In order to adapt to this new trend, MiPAAF launched the portal "Banca delle Terre" in 2017. Its role is to promote the inheritance of public land and to enhance youth entrepreneurship and generation replacement by providing ways to simplify credit and reduce land prices. ISMEA has partnered with this project and has displayed a map of the first 8,000 acres on its website.

According to the 2014–2020 Partnership Agreement, with the availability of €42 billion from EU in addition to €10. 4 billion from EAFRD, the funds for training will also be provided to Italy, which will provide an additional €21 billion for rural development measures. Therefore, in the entire mobilization meeting and plan, measures such as "transfer of knowledge and information actions" and "consultation, replacement and assistance services for management of farms" in RDP are particularly encouraged. Regions promote education through professional training courses (some of which are free of charge) in order to promote education and create added value. Some of the major themes are related to business creation and development, supply chains, and agri-food marketing, accounting and access to credit, business management, financing for agricultural enterprise, EU regulations and funds, social dimension, and organic agriculture. Through the cooperation of Coldiretti Giovani Impresa and INIPA, the training course "Impresaduepuntoterra" for young and senior graduates was launched in 2015. These training activities are promoted by ISMEA and funded by MiPAAF.

Additional master and lifelong training programs are regularly promoted within the regions through the collaboration among universities (e.g. the Food Innovation Programme of the Emilia-Romagna Region). These national activities and programs are inscribed within the European framework. Italy, as a member state, is therefore obliged to establish a Farm Advisory System (FAS) for providing advice to farmers on land and farm management. This system was introduced in 2007 as one of the major components of EU Regulation 1306/2013 CAP. The purpose of FAS is to increase farmers' awareness of material flows and farm processes related to the environment, food safety, and animal health and welfare on a voluntary basis. It also helps meet compliance requirements and avoid financial penalties.

5.1.2.5 *Evaluation Indicators and Methods of Italian CSA*

No-tillage is an important measure of CSA. It is especially useful in reducing soil machinery costs and energy consumption, machine usage fees, and working hours and can be labor-saving compared to other techniques typical of conventional agriculture. It can not only be used to measure costs and benefits but can evaluate the increase in farm efficiency and productivity. Within the Agricultural Accounting Information Network (RICA), Italy has implemented the Farm Accountancy Data Network (FADN), which aims to assess the economic status of European agriculture and to design and evaluate the CAP microeconomic agricultural level in a given area. There is a software tool called "GAIA", which has been developed by CREA-PB for users within the RICA network, agricultural entrepreneurs, consultancy services, and the world of agricultural training. The software analyses the farm business budget as a starting point through the use of specific indicators, including efficiency ones, in order to compare the technical and economic structure of a given farm to similar others. The software allows data to be collected and verified in the form of technical and economic indicators, which can be further compared. Analysis of this type of data shows that CSA solutions (especially no-tillage) can help reduce the workload per unit area of land and thus have the opportunity to use the same facilities to grow larger land. In terms of labor, compared with other sample farms that do not use this technology, the no-till "index of labor intensity" indicator has increased by 75%. Another important indicator used is the "Net Soil Productivity Index", which has a net increase of about 14% compared with other farming systems that do not use no-tillage. The Rica database used by MiPAAF and other research institutions is an effective tool for the decision-making process and for pre- and post-assessment of agricultural policies and rural development at the national level.

The climate-smart flagship project has also established another sector-specific database as a classification system that links policy incentives to specific national CSA solutions and uses the Emilia-Romagna as a case study. A similar dynamic policy knowledge database, including financial tools and policies, is intended to be an instrument for consultation by technology providers, food and beverage companies, and farmers. It represents a source of information for policymakers and comparing policies to identify gaps on climatic threats. Technology providers can combine their solutions with tools to reduce farmers' costs, and farmers can access databases to find subsidies and support for submission procedures. Therefore, subsidies and consultants can obtain information about available support through the database. Innovation policies are of high relevance in the database, such as RDP and EIP plans, water and soil regulations, national climate action plans, and regional development plans. The structure of the database is a combination of major obstacles and related policies.

5.1.3　*Forecast of Italian CSA Effects*

5.1.3.1　*The Impacts of Italian CSA on Productivity and Income and Its Implications for Food Security*

The impact assessment of CSA needs to analyze the effects of relevant policies at the beginning of the project. CSA has made a significant contribution to enhancing Italy's food security and developing food sovereignty, and estimation on productivity is a continuous process. Experts focused their calculations on the impact of CSA on soil health and crop productivity, while paying attention to the importance of CA, organic agriculture, and agroecology, as well as seed use and protection. Recent studies have paid particular attention to CA and its environmental and economic benefits, aiming to reduce greenhouse gas emissions through soil carbon sequestration and support the reduction of ammonia emissions in the field of intensive agriculture.

CA, which reduces soil operations, is expected to reap significant cost-related benefits as part of the environmental benefits by reducing fuel consumption by about 80–100 L of diesel per hectare. Each hectare of land saves €250–300 in economic costs, reducing costs, and the grain yield is higher than the soil in dry years. Soil conservation, erosion prevention, water saving, and GES contribute to biodiversity richness and fertility increase, committing especially to the Sustainable Development Goals (SDGs). Compared with traditional agriculture, CA slightly reduces crop yields but significantly saves cost input, making the ecosystem sustainable and improving soil fertility, which is extremely important for national food security and socio-economic value.

5.1.3.2 *Realization of the Adaptation Potential of Italian CSA*

Italy's adaptation policy is constantly evolving, through continuous adjustments to adapt to changing impact scenarios, to take appropriate actions to deal with the impacts of climate change. The adaptation process and actions are complex systems that require a high degree of governance and communication between the public and private sectors, as well as their responsibility centers in implementing adaptation plans across the country. In addition, the agricultural insurance system is an important economic tool, but it is applicable beforehand and is no longer suitable once damage occurs. According to the white paper on the national climate change challenge, progress and main potentialities in technological development, the adoption of agricultural system innovation, the governance, and financial management of agricultural companies (Pasqui *et al.*, 2012) has improved the ability of soil to resist erosion and drought risks in optimizing water resources management. It has also adopted technological improvements to limit agricultural water use. At the same time, it reduces nitrate content by planting crops that consume less water and reduce the impact of climate change on farmers. Sustainable and diversified use of forage crops adapted to climatic conditions is another adaptation method to increase the added value of farms. Moreover, due to the increase in temperature and considering the temperature and humidity index (THI), other adaptation measures may become more and more necessary, such as shading, aeration, and insulation systems. In addition, forecasting tools and applications for irrigation designed for end-user consultation are currently being used in order to face water scarcity or prevent floods, while securing good interventions on water volumes to be used to obtain quality products and saving water resources.

5.1.3.3 *Realization of Emission Mitigation Benefits of Italian CSA*

At the national level, the primary mitigation solutions deriving from European policies oriented toward programmes reducing emissions in the agricultural sector and Italy need to achieve binding national limitation targets for GHG emissions, as well as for the "Effort Sharing Decision" from sectors not covered by the EU Emissions Trading System (ETS). The IPCC assessment report shows that agriculture has great potential for emission reduction because it has the unique potential to act as a soil and forest carbon pool. In unprotected soil, the loss of organic matter is higher than the accumulation amount. CA can significantly reduce carbon dioxide (CO_2) emissions while consolidating and increasing organic matter reserves and contribute to the sequestration of organic matter (López-Bellido Garrido & López-Bellido, 2001). Obstacles to emission reduction exist not only in consumer awareness or consideration of economic growth but also at the socio-economic level. The loss of competitiveness and the huge economic cost of sustainable technologies are also obstacles to emission reduction. Diversified emission reduction policies include adopting strategies that adapt to emission reduction measures that affect the choice of tools to achieve predetermined goals, incentives, and commitments (increase willingness to participate in agreements voluntarily).

5.1.3.4 *Italy's CSA Economic Growth and the Realization of Other Co-Benefits*

In addition to developing resilience to climate change, reducing human pressure and strengthening management, CSA also produces many other benefits. For example, the restoration and improvement of natural habitat around development zones and farms by CSA promote the value-added of landscape and natural resources, thus promoting social and economic development. In addition to the environmental benefits of reduced pollutant presence and persistence and consequent increase in biodiversity, the local economy of many Italian regions benefits from CSA through rural tourism development and education to heritage. Through this perspective, the multifunctionality of the farm plays an important role in society. CAP believes that "agriculture plays complementarily function within society beyond its role as food producer, including the support of public service such as food security, sustainable development, environmental protection, vitality of the development of rural areas and the maintenance of a general income balance within society between farmers and other occupations".

Global analysis of the agricultural industry shows that conservation tillage increased by 47% between 2008 and 2013. CA in Italy has an area of 380,000 hectares, an increase of 450% in 5 years (FAO, 2016a). Data on CA adoption in Italy are not officially reported. In 2010, the Agricultural Census investigated the no-tillage treatment in 52,128 cultivated farms on about 290 hectares planting fields. It found that no-tillage saved about 80–100 L of diesel oil per hectare, which can bring additional economic efficiency, with saving of €250–300 per hectare. While estimations on employment rate showed that no-tillage farms in Italy employ around 210,000 operators (average 4.1 workers per farm compared to 3.8 workers per farm under conventional tillage practices) for a total of approximately 28.5 million working days per year.

5.1.4 *Opportunities and Challenges of Italy's CSA*

Italy's CSA practices are in their infancy and require time and effort to be used and promoted nationwide. When analyzing the adoption of such sustainable technologies, the issue of time is particularly sensitive. For example, in order to obtain agricultural products through conservative methods, among the arable field crops, it is necessary to wait for a soil adaptation period of about 6–8 years, during which the soil shall not be turned. Although recent studies have shown that the above practices and policies have environmental and economic values, the use of integrated systems such as traditional agriculture requires less effort from the agricultural sector.

Cost-related challenges are particularly obvious in remote areas and rural areas, where are undergoing transformation, but need more assistance in facing the associated risks. Analysis and monitoring of ongoing projects, such as those funded by the Emilia-Romagna region in 2016, provide concrete examples of experience capitalization, support for knowledge sharing and best practices, and development of new solutions and strategies. Thanks to the support of public resources at the national and local levels, the entrepreneur training recommendations can also be continuously adopted. Access to funding, technical assistance during the transition period, structural adjustments, and innovative adoption are critical to promoting the development of CSA. Support to other new projects in the current national plan is still important. These innovative projects are keen to involve farmers in establishing best practices, products, processes, and innovative technologies.

Training and mobility should be enhanced among agricultural companies, agro-industry, organizations, and the scientific community in order to overcome critical issues and develop participatory approaches, which include farmers in decision making and strategizing. Finally, attention should be paid to risk management, as the full adoption of new techniques is often perceived as an economic threat to overcome, which is also linked to criticism and concerns about CSA, which leads to limited willingness to build partnerships and collaboration. It is necessary to reduce the possibility of interest scarcity, especially at the local level, and to establish mechanisms to promote stakeholder participation and strengthen dialogue among different actors.

5.2 Mozambique: A CSA Case in Developing Countries

Mozambique is one of the most undeveloped countries located in in southeastern Africa, for which agriculture is a vital part of nation. Mozambique's agricultural production conditions are extremely backward, and production is extremely vulnerable to the natural environment. Therefore, climate change has very significant impacts on its agricultural production. Especially the increasing extreme weather events (e.g. heat stress and drought), which brings great uncertainty and great obstacles to the country's agricultural production. Faced with the adverse effects of climate change, the Mozambican government has adopted a series of CSA technologies such as CA, agroforestry management, and organic agriculture. However, these measures have not been widely adopted because of limitations on actual conditions in Mozambique. Although there are significant obstacles, there are still many institutions in Mozambique engaged in

CSA. At the same time, the country has maintained close exchanges with many national organizations on the development and research of CSA. Mozambique has also tried to introduce some policies to promote the development of CSA. Despite many efforts, there are still significant problems in the development of CSA, and the development is still relatively slow. In the future, there are still many problems to be solved in the development of CSA in Mozambique (CIAT & World Bank, 2017).

5.2.1 *The Impacts of Climate Change on Agriculture and Food Security in Mozambique*

Mozambique has a total land area of about 800,000 square kilometers with rich natural resources, about 50 million hectares of land that can be used for agricultural production. Mozambique has approximately 28 million people, 68% of whom live in rural areas. Despite the steady economic growth over the last two decades, more than two-thirds of the people still live on less than $1.9 per day and 55% of people live below the national poverty line. Mozambique was one of the seven countries to achieve the Millennium Development Goal of halving the proportion of people suffering from hunger and undernourishment by 2015. The prevalence of undernourishment declined from 56% in 1990–1992 to 24% in 2014–2016. However, the Global Food Security Index (GFSI) places the country at the lower end of the rank (108 out of 113 countries), with particularly low scores on food quality, safety, and affordability. The frequent occurrence of extreme weather events such as droughts and floods has exacerbated the challenges to food security and the pressure on survival in Mozambique.

Mozambique loses 220,000 hectares of natural forest every year, with deforestation being largely driven by reliance on fuelwood for domestic energy, as well as expansion of land for agriculture. It is estimated that 62% of the land has the potential to be used as planting fields, but only 7% land is currently cultivated. Within the 3 million hectares of land with potential for irrigated agriculture, only 118,000 hectares of land were equipped for irrigation in 2015 and just 62,000 hectares of this were being utilized for agriculture. Rice, maize, and cassava are the largest three crops, while livestock accounts for 55% of the total cultivated area. The main cash crops include sugarcane, tobacco, and cotton, which are mainly grown and processed by large multinational or state-owned companies. Mozambique has great potential to increase productivity and increase crop yields. In the past 20 years, due to the improvement of production practices and the government's vigorous promotion of sugarcane, the productivity of sugarcane has increased significantly. The livestock industry is small in scale, but it plays an important role in their national life, food security, and nutrition. Mozambique's greenhouse gas emissions are estimated at 66.72 megatons of carbon dioxide (CO_2) equivalent, and the land-use change and forestry (LUCF) sector accounts for 59% of the total greenhouse gas emissions. Agriculture is the second largest contributor, accounting for 27% of the greenhouse gas emissions, mainly through crop production in savanna, enteric fermentation, and manure left on pastures for livestock rearing. Mozambique is one of the countries in the world where climate disasters, such as droughts, floods, and cyclones, occur most frequently, and it is also one of the most vulnerable countries to climate change.

Agriculture production in Mozambique is charactered by poor small-scale farmers, who mainly produce for their livelihoods. Most of these farmers lack access to productive assets, agricultural financing, new technologies, and markets. Mozambique is working to reform the agricultural sector and attract foreign direct investments, which may help the country increase productivity and income. Limited mechanization and the use of traditional agricultural tools also hinder the development of agricultural productivity. Most farmers still use hand hoes or bull-pull tools in daily agricultural activities. Low utilization of fertilizers, pesticides, improved varieties, and other purchasing inputs is due to high cost, low availability of inputs in local markets, and lack of understanding of their use and benefits. Fertilizer, pesticide, and irrigation treatments are concentrated in the central and southern regions, and nearly 90% of the fertilizer is used for sugarcane production. Existing irrigation infrastructure and irrigable land are insufficient. Irrigation infrastructure is mainly used for foreign commercial sugarcane and other export cash crop farms. Eighty percent of fields in Mozambique rely on precipitation, which is vulnerable to climate change and related extreme weather events. Even worse, the road network in the country is poor, and there is a lack of developed input and output markets. Only 20% (around 6,000 km) of the country's national roads are paved, hindering market access and increasing agricultural transportation costs. Besides, agricultural extension services are also lagging behind. Related surveys have shown that less than 15% of farmers have accessed formal extension services annually between 2000 and 2008, and only 8.6% in 2009–2010. There are currently 2,875 public and private extension workers in this country, far below the optimal number for effective extension services. This impairs farmers' access to technical advisory on new technologies and practices implementation, which could otherwise significantly improve productivity. Furthermore, agricultural credit service access is low, particularly in rural farming communities, hindering investments in new technologies, or the maintenance of capital-intensive practices.

The biggest challenge facing the agricultural sector is extreme weather events caused by climate change, especially floods caused by droughts and hurricanes. Mozambique is one of the most vulnerable and least prepared countries with regard to natural disasters, ranking 153 out of 178 nations on the Notre Dame Global Adaptation Index (ND-GAIN). The vulnerability in this country is driven by an array of biophysical, climatic, and socio-economic factors. From 1996 to 2015, the economic losses caused by climate disasters such as droughts, floods, and hurricanes were approximately $790 million. Mozambique's coastline, which extends more than 2,700 kilometers and where half of the country's population lives, is affected by tropical cyclones, which occur at least once a year. In 2000, hurricanes caused floods, resulting in economic losses of about 20% of GDP. An estimate of drought and flood costs in 2009 indicated the average annual losses of maize and sorghum are 9% and 7%, respectively. Changes in weather and climate are also visible in the form of sea-level rise (inundation), increased incidence of wildfires, increases in mean annual temperature, increase in number of hot days, upsurge of crop and livestock pests and diseases, decreases in rainfall amounts, and shifts in seasons. Climate projections for it indicated an expected change in mean annual temperature by up to +1.4°C by 2030 and +2.2°C by 2070, with the northeast experiencing the highest increase. The greatest increases in temperatures are expected to occur between December and May. Total precipitation is not likely to decrease significantly, ranging from a 4% reduction in the north-eastern parts of the country to just 1% in the southern parts. Negative impacts of climate change on agriculture are primarily caused by the increased likelihood of extreme events such as cyclones and flooding (FAO, 2015).

5.2.2 *Technology and Practice of CSA in Mozambique*

There are hundreds of technologies and approaches around the world that fall into the category of CSA. Mozambique's National Adaptation Programme of Action (NAPA) laid a foundation for the prioritization of options to bolster farmers' resilience to climate change and identified key adaptation areas related to agriculture, water resources management, and land use. Since then, a number of CSA-related initiatives led by governmental and non-governmental actors have been implemented, including CA, agroforestry, organic agriculture, sustainable soil fertility management, and integrated pest management (IPM). Crop residue management, mulching, composting, and rotations are some of the key climate-smart practices that are being adopted across several production systems in Mozambique. Crop residue and IPM are particularly important in legume, potato, and vegetable production. The use of short season varieties and small grained crops such as sorghum have also emerged as key adaptation strategies for farmers. The country's vulnerability to drought and the underdeveloped irrigation potential has led to the uptake of many rainfall harvesting and management related CSA practices, especially for potato and vegetables cultivation. Some of the key CSA practices for livestock are rainfall harvesting, diversification of livestock breeds and species, and supplementary feeding. Given their adaptability to drought conditions, small livestock such as goats are increasingly kept by farmers as a resilience measure. A survey in 2014 showed that 52% of respondents (small-scale farmers) did not adopt CSA practices due to lack of knowledge and financial capacity to invest in on-site interventions. There is only limited quantitative evidence on the impacts of various practices on the pillars of CSA, which makes investment prioritization and implementation more difficult. Developing and revising information on the different impacts, costs, and benefits of CSA practices through field trials and monitoring systems can help build an evidence base to strengthen the appropriateness of different agricultural adaptation and mitigation methods in different regions of the country.

5.2.3 *System and Policy of CSA in Mozambique*

There are many institutions in Mozambique engaged in CSA-related activities, including the government, UN agencies, non-governmental organizations, private sector, and farmers' organizations. Their CSA-related work includes farmer capacity building through enhancing training and extension services, policy advocacy, and raising awareness. Various government ministries and departments are working on CSA, including line ministries such as the Ministry of Agriculture and Food Security (MASA), the Ministry of Land, Environment and Rural Development (MITADER), the Agriculture Research Institute of Mozambique (IIAM), the Ministry of Economy and Finance (MEF), the Council of National

Sustainable Development (CONDES), the Mozambique National Institute of Meteorology (INAM), the National Institute for Disaster Management (INGC), and the National Institute of Irrigation (INIR) among others. The Ministry of Economy and Finance — National Directorate for Monitoring and Evaluation is the country's current national designated agency to the Green Climate Fund, while the Ministry of Land, Environment and Rural Development is the coordination center of the United Nations Framework Convention on Climate Change and the Global Environment Facility. MITADER and MASA are the main government departments responsible for the promotion of rural development and agriculture. They both have mainstreamed CSA-related investments into programming, although these are not always branded as CSA.

International research institutions engaged in CSA-related research in Mozambique include the World Agroforestry (ICRAF), the International Center for Tropical Agriculture (CIAT), the International Center of Insect Physiology and Ecology (ICIPE), and the International Livestock Research Institute (ILRI). ICRAF, together with the National Directorate of Agrarian Extension (DNEA) and IIAM, actively promote agroforestry systems across the country, while ICIPE and CIAT have carried out extensive work on IPM methods and technologies and CA. Other research institutions include the International Institute for Tropical Agriculture (IITA) and the International Maize and Wheat Improvement Centre (CIMMYT), which also support CA projects. Other UN agencies include the Food and Agriculture Organization (FAO), the United Nations Development Programme (UNDP), the International Fund for Agricultural Development (IFAD), and the United Nations Environment Programme (UNEP). FAO has particularly promoted CSA, while other UN agencies have been involved in broader agricultural climate change adaptation projects and programmes. UNDP supports the development of national climate change adaptation and mitigation strategy and the national climate change monitoring and evaluation frameworks.

In Mozambique, the works related to CSA are quite uncoordinated. Although the climate change departments do exist, the establishment of a national CSA coordination mechanism would be beneficial and likely enhance the promotion and scaling up of CSA across the country. And it emphasizes the mandatory actions and investments of some key institutions directly or indirectly promoting one, two, or all of the CSA pillars. Although most of the organizations surveyed and mentioned above invest in productivity and/or adaptation activities, mitigation is seen as a co-benefit of their interventions, rather than reducing emissions. Integrating more systematic GHG emissions mitigation measures into agricultural development plans, including accounting methods and reporting, will help create a framework for more coordinated intersectoral work and contribute to the strengthening of information systems essential for CSA scale-up.

Mozambique ratified the United Nations Framework Convention on Climate Change (UNFCCC) in 1992, signed the "Kyoto Protocol" in 2005, and has so far submitted a national communication to the UNFCCC. As a least developed country, Mozambique formulated the National Adaptation Programmes of Action (NAPA) in 2007 to respond to urgent and immediate adaptation needs. These programmes have taken key actions on "strengthening the ability of agricultural producers to cope with climate change" and identified strategic areas of investments, such as increasing adaptive capacity of producers, strengthening early warning systems, and promoting the sustainable use of water resources and reduction in soil degradation especially erosion. In 2015, Mozambique submitted its Intended Nationally Determined Contributions (INDC), which recognized the contribution of forestry sector investment to climate change mitigation. However, agriculture was only identified as a priority area to adapt to NAPA, and only as a "potential" area for climate change mitigation. Mozambique was also one of the first countries to sign the Paris Agreement, and it deposited its letter of ratification at the open signing ceremony on April 22, 2016. Mozambique has also formulated the "Strategic Plan for the Development of the Agricultural Sector", "National Irrigation Strategy", and "Comprehensive Africa Agricultural Development Program Compact". The Strategic Plan for the Development of the Agricultural Sector aims to improve the competitiveness of the agricultural sector through improving use and management of natural resources (land, water, and forests) and enhancing food security and productivity. The National Agricultural Investment Plan (PNISA) in 2014 was meant to operationalize PEDSA and re-emphasize the goal of increasing productivity of major food crops, reducing hungry people by half and cutting malnutrition rate. An integral part of PNISA is natural resource management, which is a key entry point for the implementation and promotion of CSA practices.

The National Strategy for Adaptation and Mitigation of Climate Change includes objectives related to climate change adaptation and mitigation in agriculture, water and forestry management, biodiversity conservation, and social protection. The NAPA is planned to serve as the basis for a revised National Adaptation Plan (NAP) that will mainstream adaptation actions across various sectors over the next years (2015–2030). The National Irrigation Strategy aims to double the area of irrigated land, thereby playing an important role in creating an enabling environment for climate-smart practices related to water harvesting and efficient irrigation systems for small-scale farmers. Mozambique's Comprehensive Africa Agricultural Development Programme (CAADP) specifically mentioned the need to promote actions to reduce greenhouse gas emissions and support communities and producers to take mitigation measures and adapt to climate change. This may be a key entry point for large-scale promotion of CSA investment, as the CAADP is an important resource mobilization tool for the country.

In general, despite the lack of policies and strategic plans specifically aimed at promoting and expanding CSA, there is a significant body of sectoral (agriculture, environment, and forestry) policies that may create a favorable environment and become the entry point for CSA-related work in the country. However, this requires further inter-sectoral dialogue and institutional and human capacity to support the programming and on-farm implementation of CSA interventions, ensuring synergies and avoiding the duplication of efforts. These policies, strategies, and programs are regarded as the country's key CSA entry point.

5.3 Climate-Smart Crop Production System

5.3.1 *Technical System for Climate-Smart Crop Production*

5.3.1.1 *Improvement and Innovation of Crop Varieties*

Crop varieties with good adaptability are indispensable conditions for the production of climate-smart crops (FAO, 2011). National, regional, and international plant breeding works usually involve multilocational trials locations to develop crop varieties that are resistant to climate-related phenomena and more efficient in their use of resources to reduce their impact on the agricultural ecosystem and the wider environment. Drought resistance, salt tolerance, and waterlogging tolerance are the most common climate-related traits of crop varieties. Other more geographically specific factors also need attention, including higher frequency of frost during the seedling and pollination period, high temperatures during the filling period, heavy rains that compress the soil, and alternating light rain and high temperature that stimulate seed germination but prevent the establishment of seedlings.

The protection of plant genetic resources is essential for the improvement of crop varieties. They are the "raw materials" for improving and cultivating new germplasm resistant to stress. In order to deal with the new challenges brought about by climate change, it is more and more urgent to invest more resources and efforts to increase the diversity of germplasm resources. The diversity of wild relatives of crops is an important source of improved genetic traits of crops. However, due to climate change, their natural habitat is lost and their diversity may be eroded. In order to adapt the production system to climate change, breeders must develop an increasingly diverse mix of crop varieties. New varieties usually depend on obtaining heritable variation, especially from unsuitable materials, including wild relatives of crops, which are usually not used by breeders. This will involve institutionalizing and improving capacities for pre-breeding activities in which germplasm curators and breeders work together to identify the carriers of desirable traits and evaluate these putative parents and cross promising ones with elite lines to generate intermediate breeding materials. It may also involve the use of some induced mutations and other biotechnologies, such as genetic engineering and genome editing technologies. High-throughput genotyping and phenotyping platforms are increasingly being used to improve the efficiency of crop breeding. In farmer participatory plant breeding, farmers and plant breeders cooperate in the development and renewal of crop varieties. Since farmers' views help determine which varieties are proposed for official release and registration, their participation in plant breeding is an effective way to achieve demand-driven crop germplasm improvement to adapt to climate change, especially in developing countries.

Climate change will affect farmers' ability to obtain high-quality seeds and planting materials, which will jeopardize the success of subsequent cropping seasons. Seed security assessment is a method to determine seed availability, farmers' availability, and their quality and compatibility with farmers' variety preferences and production systems. This is the most appropriate countermeasure determined without hindering the development of the seed sector. The seed security assessment considers the formal and informal sources of seeds and the operation of the entire value chain to determine the main constraints that farmers face in obtaining the seeds they need. The results of the assessment can guide the next actions, which may be immediate interventions, such as direct seed distribution to farmers, support to seed markets, cash transfers, or long-term developmental activities. Two useful tools to ensure that farmers have access to high-quality seeds after a crisis in emergencies are technical handbook and practitioners' guide for seed security assessment.

5.3.1.2 *Construction of Diversified Planting Systems*

The cultivation of major grain crops such as maize, wheat, and rice often require large amounts of pesticides and herbicides. The high intensification of crop production systems leads to a single planting structure, which impairs their ability to respond to changes and provide ecosystem services (Folke, 2006). In the cropping system, greater diversity of crops and other living organisms is an important criterion for ensuring farm resilience, economic stability, and profitability. Crop diversity is particularly important in CSA because it helps control the pests and diseases and has a direct impact on yield and income. The combination of crop varieties with different genotypes is an effective means to enhance the adaptability of the production system and has a higher adaptability to climate change (FAO, 2011). When faced with abiotic changes (e.g. changes in precipitation and temperature patterns) and biotic disturbances (e.g. pest infestations), the level of existing biodiversity (functional diversity and response diversity) can make the difference between a stressed agricultural ecosystem and a resilient one. Enhancing on-farm biodiversity and integrating production also provides other environmental services, which are essential to farmers and society as a whole. The level of biodiversity of the agricultural ecosystem affects the interaction between plants, animals, and microorganisms at the landscape scale. At the state level, increasing agricultural biodiversity could effectively improve sustainable function of landscape and diets. Besides, it has huge potential on shaping rural and urban (city-region) food systems, and it could ensure the future food and nutrition security of expanding urban populations.

Achieving diversification of crop systems can take many forms, including different crop species and/or varieties (intra-specific and/or inter-specific diversification), different spatial scales (landscapes, farms, individual fields, and or crops) and different time structures. There are several ways to enhance the genetic diversity of crops such as the following:

(1) Different crop varieties of the same species can be grown in mixtures as one crop (variety mixture). For example, planting a combination of varieties that have the same growth length that can be planted and harvested at the same time, but have different responses to different water regimes, is a strategy to cope with the unpredictable start of the rainy season and improve yield stability.
(2) Different crop species can be grown together, such as relay cropping, intercropping, and crop rotations.

5.3.1.3 *Climate-Smart Soil Management*

At the field level, increasing productivity allows to grow more from the land already under production. This reduces the demand of increasing planting area and helps reduce the greenhouse gases emission associated with agricultural expansion. Improving farmland management measures could increase crop productivity and resource utilization efficiency so as to respond to and mitigate climate change. The most cost-effective management strategies for sustainable crop production include achieving a balanced cycling of nutrients through the production system and protecting the soil on the field.

Crop rotation systems that target carbon sequestration tend to maintain a positive nitrogen balance and carbon accumulation. The goal of sequestering carbon in the soil received renewed attention at the United Nations conference on climate change held in Paris in 2015 through the 'Four Per Mille' initiative. This was launched with the objective of increasing the existing soil organic carbon by 0.4% each year globally as a compensation for the global emissions of greenhouse gases. Lal (2004a) estimated that the world's cropland has the potential to store 0.4–1.2 billion tons of carbon equivalents per year. Crop residues with an average carbon-to-nitrogen ratio in range of 25–30 can be achieved by rotating between crops high in carbon and crops high in nitrogen. This allows the carbon to accumulate in the soil and enables the nitrogen in the decaying surface residues to be released slowly to the next crop. If the amount of nitrogen in the crop residues is too low, microorganisms use the mineral nitrogen existing in the soil, which reduces the amount of nitrogen available to the growing crop until the carbon in the crop residues starts to deplete (Gál *et al.*, 2007). A reasonable leguminous crop rotation system helps to sequester soil carbon, and legumes produce a large number of secretions, which is conducive to symbiosis with mycorrhiza. The decomposition of the root system of the previous crop increases the organic matter at greater depths. The deep root system is an ideal choice for bringing carbon deep into the soil, where it is less susceptible to oxidation. In agricultural ecosystems, about 80% of biological nitrogen fixation is achieved through the symbiotic association between legumes and the soil bacteria Rhizobia. Agricultural production management measures have affected these natural processes by selecting legume species that are particularly effective at fixing nitrogen. Helpful measures are increasing the proportion of legume and grass seed in forage mixtures, inoculating the legumes with bacteria (e.g. Rhizobia), improving crop nutrition (especially nitrogen and phosphorous), managing diseases and pests, choosing the best planting time, cropping sequence and cropping intensity, and managing the defoliation frequency of forage swards.

Using the best management practices for nitrogen fertilization minimizes residual soil nitrate, which reduces nitrous oxide emissions. The best management practices for nitrogen fertilization include integrated nutrient management and targeted applications of the precise amount of mineral fertilizer required. Planting crops that produce a lot of root biomass can prevent the soil from becoming compacted and improve drainage.

Crop residues can increase the content of organic matter in soil. Keeping the soil covered with a layer of evenly distributed crop residues with an average carbon-to-nitrogen ratio in the 25–30 range after harvest produces a positive residual fertilizer effect on subsequent crops. Removal of crop residues will cause the roots of crops to enter the soil organic matter pool, resulting in a decrease in the accumulation of soil organic carbon. For the same reasons, grains and legumes should be harvested by cutting plants instead of being uprooted. Mixing crop residues with soil may accelerate the fixation of nutrients in the soil and make it unavailable for subsequent crops early in the growing season. The use of machinery to mix crop residues into the soil decomposes faster than the residues left on the soil surface, and nitrogen sequestration process occurs earlier. Incorporating crop residues rich in easily decomposable carbon into the soil usually has a stimulating effect on soil organic matter and increases carbon dioxide emissions. On the contrary, when the crop residue is not mixed into the soil, its composition will not affect the decay of the stable soil organic matter already present in the soil (Kuzyakov *et al.*, 2000; Fontaine *et al.*, 2004; Sisti *et al.*, 2004).

CA is a combination of keeping soil disturbance to minimum, maintaining soil cover and diversifying crop production. It aims at reducing soil erosion and restoring degraded soil, but it provides a strategic entry point for adapting to climate change. CA strives to rebuild the most stable soil ecosystem in each agricultural ecosystem to reduce producers' dependence on external inputs for plant nutrition and pest management. Maintaining soil cover reduces water loss, stabilizes soil temperature, reduces water and wind erosion, restores soil carbon through the decomposition of crop residues, and provides food for beneficial soil organisms. Rotation and diversified crops reduce crop diseases and insect pests and supplement soil nutrients. Avoiding mechanical farming will increase the number of earthworms, millipedes, mites, and other animals in the soil. This micro-animal activity creates soil pores, improves soil structure, and partially replaces soil farming. CA combines the organic matter on the soil surface. The excrement of soil organic matter provides stable soil aggregates, and the vertical channels formed by worms drain excess water. The organic matter contained in soil microorganisms improves soil structure and water storage capacity, which in turn helps plants survive longer periods of time. Since uncultivated soil can act as a carbon sink by absorbing and storing carbon, CA is also recognized for its ability to mitigate climate change. No-tillage significantly reduces the number of farm operations required for crop production, which lowers fuel consumption (Lal, 2003). CA can bring significant energy and fuel savings, which is the reason that it has become one of the more attractive options for farmers during periods of high energy costs (FAO, 2000). Currently, CA has been adopted in approximately 120–220 million hectares of land worldwide (Prestele *et al.*, 2018).

Minimizing mechanical soil disturbance is a long-term treatment method to increase soil carbon storage. However, the accumulation of soil organic carbon is a reversible process, and any short-term disturbance, such as the periodic tillage of land under no-tillage conditions, will not lead to a significant increase in soil organic carbon (Jarecki and Lal, 2003; Fontaine, 2007; Al-Kaisi *et al.*, 2008). Although the benefits and reduced risks and costs in the future gained from improving soil health and increasing soil organic carbon accrue slowly over decades, taking action can still bring immediate financial dividends to help maintain crop productivity. When the soil is rebuilt, it will be improved and store more organic matter and water, thereby improving ecosystem functions and services (e.g. controlling rainfall runoff and soil erosion) that are vital to climate change adaptation and mitigation.

GHG emissions can be reduced by improving the efficiency of farm machinery in terms of its work and fuel usage efficiency (Choudhary *et al.*, 2002). This can be done by using the most suitable equipment for the specific farm type. Appropriate machinery, such as two-wheel tractors, combined with agronomic innovations, such as no-tillage direct seeding technology, can help adapt to and mitigate climate change. Small tractors using tined equipment instead of disk ploughs, or modern direct seeding equipment are more productive than tillage-based systems. Small equipment may be more readily available and affordable to smallholder farmers, farmer organizations, and service providers. In a mechanized system, the trade-off between improving agricultural productivity (including energy efficiency) and affordability is particularly important for the rural poor. Poor farmers are particularly desperate to increase yield and efficiency, but they are less likely to make required improvements and pay for these after they got needed resources.

5.3.1.4 *Climate-Smart Irrigation*

Irrigation plays an important role in global food security, allowing 40% of crops to grow on 20% of the world's arable land. Irrigation is the major single user of water, accounting for about 70% of the water extracted from surface water and groundwater resources. Irrigation faces some challenges, mainly related to its status as a key resource for water. The expansion of irrigated areas and the shift to water-intensive food, coupled with the increasing demand for water from other sectors, has made water an increasingly scarce resource in many parts of the world. In the future, under climate change, rainfall will continue to decrease and climate change is expected to exacerbate the challenge of water scarcity, especially in arid regions. The rising temperatures will increase crop evapotranspiration, thereby increasing the demand for irrigation water. At the same time, irrigation also exacerbates climate change, such as fossil fuel-driven pumping, the intensive use of fossil fertilizers, and GHG emissions from fossil fuel-driven machinery and automation used at all stages from crop planting to the final stage of the value chain.

A comprehensive and coordinated approach is needed to address these interrelated challenges, supporting the transition to productive and profitable irrigation systems, with resilient and well-adapted climate change, minimizing greenhouse gas emissions, and ensuring the sustainable use of water resources. Such transformation will help achieve sustainable development goals, ensuring that sustainable water resources are available to all, achieving food security and improving nutrition, and promoting sustainable agriculture. Urgent actions should be taken to address climate change and its effects. When water is a limiting factor, water management can be improved by protecting soil and water resources, and then crop yield per unit water resource can be maximized by reducing deficit irrigation, and unproductive evaporation loss with more effective irrigation technology. Usually increasing irrigation efficiency requires additional energy costs, so it is of great significance to adopt appropriate energy technologies when expanding irrigation. Strategies to change the management and governance of agricultural water use must be completed by incorporating water balance analysis into the decision-making process, and water balance assessments are conducted at both field and water levels.

5.3.1.5 *Integrated Pest Management*

Climate change will affect the occurrence and spread of various diseases, pests, and weeds. This phenomenon is mainly the result of changes of the distribution of host plants and crops, and natural enemies, and also of their health changes and the adaptive changes in farm management. With the increasing globalization of the trade and exchange of germplasm, these changes will exacerbate the challenges of pest management. Integrated pest management is an ecosystem-based crop production and protection method, which is based on careful consideration of all existing pest management techniques. Integrated pest management includes taking appropriate measures to prevent the development of pest populations, maintaining pesticides and other interventions at an economically reasonable level; reducing or minimizing risks to human health and the environment; and destroying agricultural ecosystems as little as possible. The ability to make correct decisions in the wild is essential for effective comprehensive pest management. The principles of FAO integrated pest management approach includes growing healthy crops, understanding the field's ecological processes, and encouraging natural pest management mechanisms that maintain ecological balances among populations of pests and their natural enemies (predators, parasitoids, antagonists), observing fields regularly, building farmers' capacity and understanding of ecological needs so that they are empowered to take the best pest management decisions in their own fields. Multi-crop systems can play a role in enhancing the resilience of crop systems and providing ecological insurance to prevent crop failures. Reasonable farming measures also contribute to the prevention and control of certain pests.

Integrated pest management is valid in a variety of different and evolving farming conditions. As climate change intensifies, national regulations, policies, and institutional frameworks must be strengthened to enable farms and rural communities to adopt integrated pest management methods. In particular, CSA frameworks should support farmer training in integrated pest management; maintain the surveillance systems, including those used in community groups that are used to detect and report changes in the behaviors of pests and natural enemies and develop appropriate quarantine procedures to prevent the entry and establishment of plant pests and formulate appropriate management strategies to respond to potential outbreaks. Strategies to promote the transition to an adaptive crop production system also include other measures. Phytosanitary frameworks and measures that can facilitate the creation of markets for sustainable products, and the transparent collaboration among policymakers, industries, and farmers on the national registration processes for the most appropriate pesticides to a climate-smart approach (FAO & INRA, 2016). Regionally and internationally, common regulations and strategic frameworks, including the International Plant Protection Convention (IPPC) and the FAO International Code of Conduct for Pesticide Management, are used to limit the impact of invasive species and the unregulated use of chemical pesticides. However, the pest management system will benefit from more coordinated actions to prevent the crises associated with transboundary pests, major pest outbreaks, and climate change. By establishing new partnerships and alliances, we can meet this common challenge at the global, national, and regional scales (Allara *et al.*, 2012).

5.3.1.6 *Agricultural Mechanization*

Improving the level of agricultural mechanization is an important way to implement sustainable crop management and increase the productivity of each land unit. Agricultural mechanization is conducive to creating better sowing and harvesting conditions, and it is an effective means to deal with climate change. It also increases efficiency in the various production and processing operations and in the production, extraction, and transport of agricultural inputs, including coal and oil. Investment in mechanization enables farmers to expand their scope of activities and reduce the impact of climate change on farmers' livelihoods. Sustainable mechanization can create opportunities to provide rental services for field operations, improve transportation and processing of agricultural products, and increase the possibility of value-added agricultural production. From a long-term perspective, the initial investment in mechanization will be compensated in the next few years, that is, increasing the rate of return of agriculture and labor, achieving surplus production or increasing the amount of production land, and improving resource utilization efficiency and related savings. Examples of the proper use of agricultural machinery in common agricultural production are listed:

(1) Using smaller tractors, making fewer passes across the field, and reducing working hours, when combined with CA, reduce carbon dioxide emissions, minimize soil disturbance, and curtail soil erosion and degradation that are common in tillage-based crop systems.

(2) Tractor-operated tillage is the most energy-consuming operation type. CA is flexible enough to accommodate the socio-economic resources of small farmers and large-scale agricultural operations. Minimal soil disturbance can be achieved by digging rods, jab planters, or mechanized direct seeders, which are specially developed to drill the seed through a vegetative layer.

(3) With the introduction of CA, the machinery park of mechanized farms has become equipment that requires less pulling power than a plough does. This means that smaller tractors, including two-wheeled tractors, can be used. Besides, it also means less fuel consumption, reduced working hours, and reduced equipment depreciation. All these will lead to a reduction in emissions from various farm operations and machinery manufacturing (Lal, 2016).

(4) The use of suitable supporting machinery enables producers to efficiently plant, harvest, and process crops, which will help increase production and reduce post-harvest losses. The case shows that small farmers in Zambia have increased their chances of obtaining sustainable mechanization technologies by improving the CA mechanization.

(5) Precision agricultural equipment along with controlled release and deep placement technologies make it possible to accurately match

production inputs with plant needs. This improves fertilizer efficiency and reduces GHS emissions caused by fertilizer application.

5.3.1.7 *Weather Forecast*

Reasonable and scientific agrometeorological forecasting is an important support for guiding farmers' decision-making and formulating strategies to deal with climate change. Weather forecasting and early warning systems are important technologies for risk management. The improved timing and reliability of seasonal forecasts and hydrological monitoring enables farmers to make better use of climate information, take pre-emptive actions, and minimize the impact of extreme events (Faurès *et al.*, 2010; Gommes *et al.*, 2010). In modern commercial horticultural production systems, weather stations often monitor irrigation conditions based on crop water requirements. In this way, irrigation systems can automatically adjust to climate change, and information and communication technologies can also support the exchange of information needed to address climate change adequately.

5.3.2 *Priorities and Policy Requirements for Developing Climate-Smart Crop Production Technology*

5.3.2.1 *Technical Research Priorities for Climate-Smart Crop Production*

Responding to future challenges related to climate change requires more research investments, including building the evidence base for climate-smart interventions and technologies, tailoring the strategies that have proven to be effective to increase their applicability in specific locations, and accelerating the development and adoption of new promising technologies and practices. Current crop research is focused on cereals (maize, wheat, and rice) and legumes (soybeans and peanuts). Maintaining the health of the agricultural ecosystem under different climatic conditions requires increasing the biodiversity of the production system and constructing more reasonable intercropping and rotation models. Expanding the scope of research crops would increase farmers' adaptation options (Glover *et al.*, 2010). The development of new crop varieties and commercially sustainable perennial grain crops varieties to resist drought, floods, salinity, diseases, and insect pests will involve the protection of genetic pools of multiple varieties, land races, rare breeds, and closely related wild relatives of domesticated species to maintain a genetic bank for the selection of novel traits.

Other research priorities in climate-smart crop production include the investigation of methods for adapting farming practices and technologies to site-specific conditions and needs. The adoption of climate-smart technologies is not simply determined by directly assessing the adaptability and resilience of a single crop to a specific environment. The suitability of any intervention requires a comprehensive scientific investigation to assess the constraints of farmers' difficulty in adopting climate-smart crop production systems. Climate-smart planting systems must be designed, tested, and validated, and farmers must participate in identifying barriers to adopting these systems and developing strategies to overcome them. However, in many countries, research on crops, soils, and water are separated into different institutions, and they are given different priorities. This fragmentation of research efforts is a major constraint for the efficient and integrated management of crops, soil, water, and nutrients, and ultimately hinders the transition to CSA. Promoting and supporting integrated research produces important public goods.

The exchange of research outputs must be more policy-friendly. Researchers need to provide decision makers and development practitioners with clear key information and provide them with the tools they need to prioritize potential policies and strategies. To foster the uptake of research by producers, innovative agricultural system approach should be adopted widely and research priorities should be determined by practical experience.

5.3.2.2 *Policy Needs for Climate-Smart Crop Production*

To formulate and implement local and effective climate change adaptation and mitigation strategies for crop production, it is necessary to strengthen scientific and technological capabilities at many levels to create a favorable environment for change and to enhance the capabilities of all groups related to climate intelligence in crop production. The main stakeholders include national researchers, policymakers, extension agencies, farmers, and the private sector, especially small and medium-sized enterprises. It is not only agriculture and agriculture-related enterprises that need support in acquiring skills and knowledge but also people who are not familiar with agricultural production. The capabilities of policymakers, extension agencies, agricultural entrepreneurs, and farmers need to be continuously strengthened and updated. This requires strengthening organizational and institutional capabilities, such as coordination mechanisms. It is recommended to adopt a system-wide capacity development method, because the production of climate-smart crops is knowledge-intensive, highly location-specific, and closely related to global dynamics.

It is particularly important for farmers in gaining and sharing knowledge about changing climatic conditions and the sustained viability of adapted crop production practices when formulating strategies to cope with the limiting factors affecting their crop system. They could help better allocate the resources they have at their disposal and those they can mobilize and make reasonable investments in climate change adaptation and/or mitigation. It is important to understand the process that farmers go through when deciding to adopt new practices and technologies. This is only possible at the local level. Diversified and demand-driven extension services play a key role in facilitating actual changes in climate-smart crop production. These services provide opportunities to gain and share technical experience and improve farmers' ability to implement these measures. They also help reduce the risk of failure caused by the transition to new systems and new business methods. For example, the Farmer Field School provides a local platform for collaboration among farmers, extension agencies and researchers, and helping to develop regional strategies to adapt to climate change. These programs often combine capacity development at the local level with broader policy frameworks and governance-related actions. In many countries, public extension services have deteriorated. Some of them have been replaced by various entities (e.g. research institutions, government ministries, farmers' organizations) via mobile phones, the Internet, radio, and television. Individual input providers and service providers (e.g. throughout-grower schemes) have also increased. Therefore, many farmers, especially women, cannot obtain any form of extension services. The needs of women farmers must not be ignored because they are important as food producers in many countries.

Improving local ability to select and evaluate crop varieties is essential to ensure that farmers can obtain locally suitable varieties. This requires the creation of a platform for participatory variety breeding and evaluation at the community level. FAO has developed a multi-module toolkit to support capacity building across the entire seed value chain, including production, processing, quality assurance, and marketing by small and medium enterprises.

The development of private sector capacity in manufacturing and providing services and selling agricultural machinery can also support the adoption of climate-smart crop production practices. In most developing countries, the lack of locally manufactured agricultural machinery and spare parts, as well as the lack of local repair and maintenance services, are important obstacles to achieve sustainable mechanization and lead to inefficient crop production.

5.3.3 *Practical Cases of Climate-Smart Crop Production System*

5.3.3.1 *Crop Rotation Systems in Europe*

Europe is an important agricultural production base in the world, with a highly developed agricultural production. European agriculture has successfully transformed from highly intensive agriculture, and its exploration in CSA practices is worth learning. The agricultural production in Denmark is a typical case of the friendly development of agricultural production and environmental protection. Denmark is one of the most intensively farmed regions in the world, where more than 60% of its land area is used for agriculture, whose food exports are more than twice of the national consumption (FAO, 2014), and it has one of the world's most well-developed environmental regulation systems (van Grinsven *et al.*, 2012).

For centuries, Denmark has been the main food supplier to the surrounding countries. It is the biggest beef and grain supplier for Germany and Norway, respectively. It is also the butter and bacon supplier for the UK and other overseas countries. With the introduction of cooperative dairy and slaughter houses in the late 1800s (Odgaard and Rømer, 2009), initially, the increase in food production was driven by the biological nitrogen fixation of grass-clover and pulses and the significant expansion of the proportion of cultivated land (Kjaergaard, 1994; Dalgaard and Kyllingsbæk, 2003; Dam and Jakobsen, 2008). However, the greatest expansion of food production occurred after World War II and it was driven by synthetic N-fertilizer inputs, increasing nitrogen content from 15 kg N hm^{-2} in 1945 to 143 kg N hm^{-2} in 1983 (Dalgaard and Kyllingsbæk, 2003; Dalgaard *et al.*, 2005), and falling to 74 kg N hm^{-2} today (Statistics Denmark, 2012).

After World War II, the expansion of agricultural nitrogen inputs gradually led to an increase in the nitrogen surplus in farmland soils and significantly increased the amount of nitrogen leached in groundwater

(Hansen *et al.*, 2011). In addition, 60% of the land area in Denmark is used as the planting field. And 7,500-km long coastline has plenty of shallow estuaries and coastal waters, which have caused serious environmental problems. According to the EU Nitrate Directive, Denmark has defined the entire territory as a nitrate-vulnerable area. Almost 100% of Denmark's drinking water comes from groundwater, and the rising concentration of nitrate in groundwater increases the risk of drinking water. In 2005, about 15% of the land was designated as a nitrate-vulnerable area (Hansen and Thorling, 2008).

Since 1985, Denmark has implemented a series of policy measures to reduce environmental pollution caused by the loss of nitrogen and other nutrients (Dalgaard *et al.*, 2005; Kronvang *et al.*, 2008; Mikkelsen *et al.*, 2010), including: (1) 1985 Action Plan on nitrogen, phosphorus, and organic matter (NPO); (2) 1987 Action Plan for the Aquatic Environment I (AP-I); (3) 1991 Action Plan for Sustainable Agriculture; (4) 1998 and 2000 Action Plans for the Aquatic Environment II (AP-II); (5) 2001 Ammonia Action Plan; (6) 2004 Action Plan for the Aquatic Environment III (AP-III); (7) 2009 Green Growth Plan; (8) 2011 Draft Plans for River Basin Management Plans (RBMPs), implementing the EU Water Framework Directive (WFD).

Already in 1972, Denmark, France, Iceland, Norway, Portugal, Spain, and Sweden signed the "Oslo Convention", prohibiting the direct dumping of harmful substances at sea. After the United Kingdom, the Netherlands, and Germany were involved, the treaty was revised in 1981 and today is included in the "Convention for the Protection of the Marine Environment of the North-East Atlantic" (1992). The Danish action plan has been subsequently used to implement the EU Nitrate Directive in 1991 (EU, 1991) and the WFD in 2000 (EU, 2000). In addition, at an international level, the ambitions of reducing nutrient loads to the environment are also important parts of treaties in relation to the HELCOM Baltic Marine Environment Protection Commission (Van der Straeten *et al.*, 2011), the Marine Strategy Framework Directive (EU, 2008), and then from 1983 and onwards enforced UN Convention on Long-range Transboundary Air Pollution ((UNECE, 1979; Sutton *et al.*, 2014). These initiatives and actions have led to a significant reduction in the soil nitrogen surplus in Danish farmland. The development of the nitrogen surplus has a good correlation with the implementation of the action plan and the concentration of nitrate in groundwater.

Denmark's effective policy measures to reduce nitrogen emissions focus on better utilization of nitrogen in fertilizers and lower fertilization nitrogen standards (Grant and Waagepetersen, 2003; Børgesen *et al.*, 2009, 2013). Denmark has implemented the statutory regulations on the minimum utilization of nitrogen in organic fertilizers and imposed strict maximum limits on the nitrogen that can be applied. This means farmers must record their application of nitrogen fertilizer according to the national standard system, document their use of manure and mineral fertilizers in the mandatory fertilizer and crop rotation plans, and adhere to the relevant provisions of the minimum replacement rate of mineral fertilizers and manure (Petersen & Sørensen, 2008; MAFF, 2013).

5.3.3.2 *Soybean–Maize Crop Rotation System in the United States*

The United States is the world's largest producer of maize and soybeans, and the planting area of maize and soybean is always equal. Some measures, such as constructing reasonable soybean-corn rotation modes, including corn-soybean, corn-corn-soybean, and corn-soybean-soybean, together with conservation tillage technology with no tillage as the core, have brought significantly positive effects in improving agricultural system productivity and carbon sequestration and emission reduction. Soil management practices are one of the main factors affecting the exchange of greenhouse gases between soil and the atmosphere that led to global climate change (IPCC, 1996). Therefore, measuring soil gas emissions from different farming treatments and crop systems is essential to determine management practices that can positively affect the carbon balance and greenhouse gas emissions (Post *et al.*, 1990; Reicosky, 1997). Maize–soybean rotation is the most important farming system in the Midwestern United States. No-tillage (NT) accounts for more than 22% of the total arable land area, while traditional farming systems account for more than 35% (CTIC, 2004). Historically, most fields in the Midwest that were not in NT or MP were CP. About 2 million hectares of poorly drained farmland in the Midwest must be completely drained of water (USDA-NRCS, 1987) and improvement of drainage is conducive to the adoption of conservation tillage. Farming measures and crop rotation have a significant impact on CO_2 and CH_4 emissions, but the crop rotation system has no impact on CO_2 and CH_4 emissions in the short term after cultivation. The percentage of changes explained by temperature and humidity is less than 27% and varies with the farming system. However, the soil temperature and humidity conditions of the 10 cm surface layer are significantly related to CO_2 emissions. Normally, the CO_2 emissions of continuous cropping maize are 16% higher than that of maize in the rotation system. After 30 years of continuous cultivation and rotation measures, the yields of maize and soybean in these areas have been higher than the average level in the Midwest. And long-term maize and soybean rotation under NT measures is considered the least carbon emission system accounting to the evaluation of the maize growth process.

5.3.3.3 *Rice Production Systems*

More than 90% of the rice in the world is produced in flooded fields. Rice grows in continuously flooded rice fields. Although the transpiration rates are similar, it absorbs two to three times the water of other irrigated cereals. Therefore, the production of 1 kg of paddy rice usually requires up to 2,500 L of water, which means that only 0.4 kg of rice needs 1 m^3 of water (Bouman *et al.*, 2007). Flooded rice fields are one of the main sources of methane emissions. The total amount of methane emitted from rice fields is about 625 million tons of carbon dioxide equivalent each year (FAO, 2016b). In continuously flooded fields, the drainage after the end of the growing season releases methane formed by anaerobic decomposition of organic matter. And the release of nitrogen is mainly in the form of ammonia volatilization (Xu *et al.*, 2012). In irrigated rice and lowland rice production, farmers have many ways to save water. These options include no-till in combination with mulching to provide soil cover, raised beds, and land levelling; alternate wetting and drying irrigation; and aerobic rice (Bouman *et al.*, 2007).

The aerobic rice production system uses specially developed aerobic rice varieties that grow in well-drained, water-free, and unsaturated soils. Since aerobic rice requires less water in the field than traditional rice, this system is used in relatively water-scarce irrigation or rain-fed lowland environments. Irrigation can be carried out by flood irrigation, furrow irrigation (with the rice growing on raised beds), or sprinklers. In this system, there are many types of weeds and they always growth rapidly, so the control effect is very important. In addition, soil-borne diseases such as nematodes, root aphids, and fungi are more common than in flooded rice, especially in tropical areas. Therefore, it is recommended to be used in conjunction with suitable dryland crop rotation to control rice diseases, insects, and weeds. The nutrient management strategy adapted to local conditions can effectively achieve the target yield of 4,000–6,000 kg/hm^{-2} in the rice production system.

Non-continuous water system, like alternate wetting and drying, reduce water demand and allows water to be allocated for other uses. This is particularly beneficial in the main irrigated rice areas, mainly due to the greater pressure on water supply in these areas. At the same time, alternate wetting and drying irrigation reduces power consumption and reduces the production cost of farmers. In addition, alternating wetting and drying irrigation reduces the anaerobic time of rice fields and reduces methane emissions by about 16% but may increase nitrous oxide emissions. The recurring shift between aerobic and anaerobic conditions enhances nitrification, and if the nitrates are not taken up by the plants, nitrogen may be released into the atmosphere through denitrification (the biological reduction of nitrates to nitrogen gas by bacteria). In the alternate wetting and drying technique, irrigation water is used when the rice plants mature. First, two weeks of flooding can stop the growth of weeds, and then drain away the water to dry the field until the water level drops to 15 cm below the soil surface. During flowering season, a layer of water 3–5 cm deep should be maintained. And during grain filling stage, the alternate flooding and drying scheme should be repeated until two to three weeks before harvest. Alternate wetting and drying can be applied with different rice production methods. It can be used to replace continuous flooding and also can be adopted in intensive rice production systems.

5.3.3.4 *CA in Africa*

Africa is rich in thermal resources but lacks agricultural resources and developed agricultural technology. Crop production in Africa has caused serious soil damage. Protective agricultural water storage and moisture conservation have played a key role in agricultural production in this area. Farmers in northern Namibia are adopting CA practices to grow drought-tolerant crops, including millet, sorghum, and maize. The farming system uses a soil loosening furrow machine towed by a tractor to tear the hard disk to a depth of 60 cm, forming furrows for field rainwater collection.

The collected water is concentrated in the root area of the crop and is planted in the rift line together with fertilizer. The tractor was used to build the system in the first year. From the second year, farmers used no-tillage direct seeding technology for sowing operations. Although crop stalks are mainly used as animal feed, the increase of biomass also increases the crop stalks covered by the ground to a certain extent. At the same time, CA encourages farmers to develop soybean–gramineous crops rotation mode, which can effectively improve soil structure and increase fertility and water retention capacity. After applying conservation agricultural technology, the average yield of maize has increased from 300 kg per hectare to more than 1,500 kg per hectare.

References

Al-Kaisi M, Kruse M, Sawyer J. 2008. Effect of nitrogen fertilizer application on growing season soil CO_2 emission in a Corn-Soybean rotation. *Journal of Environmental Quality*, 37: 325–332.

Allara M, Kugbei S, Dusunceli F, *et al.* 2012. Coping with changes in cropping systems: Plant pests and seeds. FAO/OECD Workshop on Building Resilience for Adaptation to Climate Change in the Agriculture Sector, 23–24 April 2012.

Bouman M, Lampayan R M, Tuong T P. 2007. Coping with water scarcity. In: Hardy B (ed). *Water Management in Irrigated Rice*. International Rice Research Institute, Los Baños, pp. 17–22. http://books.irri.org/9789712202193_content.pdf

Choudhary M, Akramkhanov A, Saggar S. 2002. Nitrous oxide emissions from a New Zealand cropped soil: Tillage effects, spatial and seasonal variability. *Agriculture Ecosystems & Environment*, 93: 33–43.

CIAT & World Bank. 2017. CSA in Mozambique. CSA country profiles for Africa series. International Center for Tropical Agriculture (CIAT); World Bank, Washington, DC, p. 25.

CTIC. 2004. *National Crop Residue Management Survey: A Survey of Tillage System Usage by Crops and Acres Planted*. West Lafayette: Conservation Technology Information Center.

Dalgaard T, Børgesen C D, Hansen J F, *et al.* 2005. How to half N-losses improve N-efficiencies and maintain yields? The Danish Case 3rd International Nitrogen Conference Contributed Papers. Z Zhu, K Minami, G Xing (eds.) (pp. 291–296). Monmouth Junction, NJ: Science Press.

Dalgaard T, Kyllingsbæk A. 2003. Developments in the nitrogen surplus and the fossil energy use in Danish agriculture during the 20th century. In J L Usó,

B C Patten, C A Brebbia (eds.) *Advances in Ecological Sciences*, Vol. 18 (pp. 669–678). Southampton: WIT Press.

Dam P, Jakobsen J G G. 2008. Historisk-Geografisk Atlas. In J Brandt, R Guttesen, P Korsgaard, *et al.* (eds.) Atlas over Danmark odense: Det Kongelige Danske Geografiske Selskab og Geografforlaget, p. 179.

Doets C E M, Best G, Friedrich T. 2000. Energy and CA. Internal FAO paper. Rome.

FAO & INRA. 2016. Innovative markets for sustainable agriculture – How innovations in market institutions encourage sustainable agriculture in developing countries. In: Loconto, A., Poisot, A.S. & Santacoloma, P. (eds.), FAO, ROME. https://www.fao.org/3/i5907e/i5907e.pdf

FAO. 2000. The Energy and Agriculture Nexus. Chapter 2: Energy for Agriculture. FAO, Rome. https://www.fao.org/3/x8054e/x8054e00.htm#P-1_0.

FAO. 2011. *Save and Grow, a Policymaker's Guide to the Sustainable Intensification of Smallholder Crop Production*. FAO, Rome.

FAO. 2014. *FAOSTAT, Food and Agriculture Organization of the United Nations Statistics Country Profile for Denmark*. Rome: FAO.

FAO. 2015. Climate change and food security: risks and responses. FAO, ROME. https://www.fao.org/3/i5188e/I5188E.pdf.

FAO. 2016a. *Global Alliance for CSA Case Study, Italian Synergies and Innovations for Scaling-up CSA*. Rome: FAO.

FAO. 2016b. The State of Food and Agriculture 2016—Climate change, agriculture and food security. FAO, ROME. http://www.fao.org/3/ai6030e.pdf

Faurès J M, Bernardi M, Gommes R. 2010. There is no such thing as an average: How farmers manage uncertainty related to climate and other factors. *International Journal of Water Resources Development*, 26 (4): 523–542.

Folke. 2006. Resilience: The emergence of a perspective for social ecological systems analyses. *Global Environmental Change*, 16: 253–267.

Fontaine S, Bardoux G, Abbadie L, *et al.* 2004. Carbon input to soil may decrease soil carbon content. *Ecology Letters*, 7: 314–320.

Fontaine S. 2007. Stability of organic carbon in deep soil layers controlled by fresh carbon supply. *Nature*, 450: 277–280.

Gál A, Vyn T J, Michéli E, *et al.* 2007. Soil carbon and nitrogen accumulation with long-term no-till versus moldboard plowing overestimated with tilled-zone sampling depths. *Soil and Tillage Research*, 96: 42–51.

Glover J D, Reganold J P, Bell L W, *et al.* 2010. Increased food and ecosystem security via perennial grains. *Science*, 328: 1638–1639.

Gommes R, Acunzo M, Baas S, *et al.* 2010. Communication approaches in applied agrometeorology. In K Stigter (eds.) *Applied Agrometeorology* (pp. 263–287). Heidelberg: Springer.

Grant R, Waagepetersen J. 2003 December. Vandmiljøplan II — Slutevaluering (aquatic action plan II — Final evaluation) Report. National Environmental Research Institute, Silkeborg, p. 32.

Hansen B, Thorling L, Dalgaard T, *et al.* 2011. Trend reversal of nitrate in Danish groundwater — A reflection of agricultural practices and nitrogen surpluses since 1950. *Environmental Science Technology*, 45: 228–234.

Hansen B, Thorling L. 2008. Use of geochemistry in groundwater vulnerability mapping in Denmark. *Geological Survey of Denmark and Greenland Bulletin*, 15: 45–48.

IPCC. 1996. Intergovernmental panel on climate change (IPCC): Impacts, adaptations and mitigation of climate change: Science technical analysis. In R T Watson, *et al.* (eds.) *Climate Change: The IPCC Scientific Assessment* (pp. 745–771). Cambridge: Cambridge University Press.

Istat. 2015. The Annual report 2014 – The state of the Nation. Rome, Italy. https://www.istat.it/it/files//2015/07/Sintesi-RA2015_En_Def.pdf.

Istat. 2016. The Annual report 2016 – The state of the Nation. Rome, Italy. https://www.istat.it/en/files/2016/07/Annual-Report-2016.pdf.

Jarecki M K, Lal R. 2003. Crop management for soil carbon sequestration. *Critical Reviews in Plant Sciences*, 22: 471–502.

Kjaergaard T. 1994. *The Danish Revolution, 1500–1800: An Ecohistorical Interpretation.* Cambridge: Cambridge University Press, p. 314.

Kronvang B, Andersen H E, Børgesen C D, *et al.* 2008. Effects of policy measures implemented in Denmark on nitrogen pollution of the aquatic environment. *Environmental Science and Policy*, 11: 144–152.

Kuzyakov Y, Friedel J K, Stahr K. 2000. Review of mechanisms and quantification of priming effects. *Soil Biol Biochemistry*, 32: 1485–1498.

Lal R. 2003. Carbon emission from farm operations. *Environment International*, 30 (7): 981–990.

Lal R. 2004a. Soil carbon sequestration impacts on global climate change and food security. *Science*, 304 (5677): 1623–1627.

Lal. 2016. Soil health and carbon management. *Food and Energy Security*, 5: 212–222. https://doi.org/10.1002/fes3.96.

López-Bellido Garrido R J, López-Bellido L. 2001. Effects of crop rotation and nitrogen fertilization on soil nitrate and wheat yield under rainfed Mediterranean conditions. *Agronomie*, 21: 509–516. https://doi.org/10.1051/agro:2001140.

MAFF. 2013. Vejledning Om gødnings-Og Harmoniregler (Guide on F and Harmonisation Regulations) (Copenhagen, Denmark: The Danish Agrifish Agency, Ministry of Food, Agriculture and Fisheries), p. 150.

Mikkelsen S, Iversen T M, Jacobsen B H, *et al.* 2010. Danmark — Europe reducing nutrient losses from intensive livestock operations. In P Gerber, H Mooney and J Dijkman (eds.) *Livestock in a Changing Landscape Vol 2: Experiences and Regional Perspectives*, Chapter 8, pp. 140–153.

MiPAAF. 2011. Sfide e opportunità dello sviluppo rurale per la mitigazione e l adattamento ai cambiamenti climatici, Libro Bianco.

Odgaard B V, Rømer J R. 2009. Danske Landbrugslandskaber Gennem 2000 år [M]. Fra Digevoldinger Til Støtteordninger, pp. 253–281.

Pasqui M, Tomozeiu R, Bucchignani E, *et al*. 2012. Libro bianco-Sfide ed opportunità dello sviluppo rurale per la mitigazione e l'adattamento ai cambiamenti climatici. Chapter: Scenari di cambiamento climatico, impatti sull'agricoltura e adattamento. In Imago Editrice SRL (eds), *Pubblicazione realizzata con il contributo FEASR (Fondo europeo per l'agricoltura e lo sviluppo rurale)*, pp. 36–86. https://www.reterurale.it/flex/cm/pages/ServeBLOB.php/L/IT/IDPagina/5799.

Petersen J, Sørensen P. 2008. Fertilizer value of nitrogen in animal manures — basis for determination of a legal substitution rate. In: Petersen J, Sørensen P. (eds), Institut for Jordbrugsproduction og Miljø, Tjele, Denmark. https://www.cabdirect.org/cabdirect/abstract/20093032474.

Post W M, Peng T H, Emmanuel W R, *et al*. 1990. The global carbon cycle. *Agronomy for Sustainable Development*, 78: 310–326.

Prestele R, Hirsch A L, Davin E L, *et al*. 2018. A spatially explicit representation of CA for application in global change studies. *Global Change Biology*, 24: 4038–4053.

Reicosky D C. 1997. Tillage-induced CO_2 emission from soil. *Nutrient Cycling Agroecosyst*, 49: 273–285.

Sisti C P J, Santos H P D, Kohhann R, *et al*. 2004. Change in carbon and nitrogen stocks in soil under 13 years of conventional or zero tillage in southern Brazil. *Soil & Tillage Research*, 76 (1): 39–58.

Statistics Denmark. 2012. Agricultural and horticultural economy in Agriculture, horticulture and forestry. https://www.dst.dk/en/Statistik/emner/erhvervsliv/landbrug-gartneri-og-skovbrug/oekonomi-for-landbrug-og-gartneri.

Sutton M A, Oenema O, Dalgaard, T. 2014. Nitrogen on the table: The influence of food choices on nitrogen emissions and the European environment. ENA Special Report on Nitrogen and Food. The Food Climate Research Network, eds. H Westhoek.

The Council of the European Communities. 1991. Council Directive of 12 December 1991 Concerning the protection of waters against pollution caused by nitrates from agricultural sources (91/676/EEC). https://eur-lex.europa.eu/legal-content/EN/TXT/?uri=CELEX%3A01991L0676-20081211.

The European Parliament and the Council of the European Union. 2000. Directive 2000/60/EC of the European Parliament and of the Council of 23 October 2000 establishing a framework for Community action in the field of water policy. https://eur-lex.europa.eu/legal-content/EN/TXT/PDF/?uri=OJ:L:2000:327:FULL.

The European Parliament and the Council. 2008. Directive 2008/56/EC of the European Parliament and of the Council of 17 June 2008 establishing a framework for community action in the field of marine environmental policy (Marine Strategy Framework Directive). https://eur-lex.europa.eu/legal-content/EN/TXT/?uri=CELEX%3A32008L0056.

The Helsinki Commission. 2008. 1992 Markets of concentration permits: The case of manure policy. Convention on the Protection of the Marine Environment of the Baltic Sea Area, p. 43.

USDA-NRCS. 1987. Farm drainage in the United States: History, status and prospects. Misc. Publ. No. 1455. USDA-NRCS, Washington, DC.

UNECE. 1979. 1979 Convention on Long-range Transboundary Air Pollution. https://unece.org/sites/default/files/2021-05/1979%20CLRTAP.e.pdf.

Van der Straeten B, Buysse J, Nolte S, *et al.* 2011. Markets of concentration permits: The case of manure policy. *Ecological Economics*, 11: 2098–2104. https://doi.org/10.1016/j.ecolecon.2011.06.007.

Xu J, Peng S, Yang S, *et al.* 2012. Ammonia volatilization losses from a rice paddy with different irrigation and nitrogen managements. *Agricultural Water Management*, 104: 184–192.

Chapter 6

Background of China's Climate-Smart Agricultural Development

Abstract

Climate-smart agriculture is an emerging concept and popular in recent years. Its core objective is to improve the ability of agricultural production to adapt to climate change while continuously increasing agricultural productivity and farmers' income, and reducing or eliminating agricultural greenhouse gas emissions as much as possible. This chapter systematically summarizes the background of the development of China's climate-smart agriculture from the perspective of the multiple challenges facing China's agriculture such as climate change, the role played by climate-smart agriculture in China's agricultural transformation and development, and the relevant foundation of China's climate-smart agriculture. It focuses on the needs of China's agricultural transformation for ensuring food security, coping with and mitigating the impacts of climate change, protecting resources and the environment, promoting farmland carbon sequestration and emission reduction, and increasing farmers' income. The chapter also analyzes the theoretical path of realizing the multidimensional goals and demands of agricultural transformation by climate-smart agriculture. At the same time, it attempts the current understanding and research of climate-smart agriculture theory in China and summarizes the supporting policies of CSA in the recent stage. The research and application progress of key climate-intelligent agricultural technologies such as conservation tillage and straw returning technology are also summed up.

135

6.1 China's Agricultural Production Is Facing Multiple Challenges Due to Climate Change

Climate change has had a great impact on the world and human survival, and the agricultural production in China and the world were hugely impacted by it. How to deal with and mitigate the adverse effects of climate change is a major challenge for China's agricultural production transformation at present and future. On the other hand, under the pressure of complex external environment and internal needs, China's food security, especially the grain security, shouldn't be ignored. In addition, the national "14th five-year plan" and the proposal for the long-term goal of 2035 call for giving priority to the development of agricultural and rural areas, and comprehensively promoting the revitalization of rural areas. Therefore, in the context of climate change, China's future agricultural production will face multiple challenges, such as ensuring food security, adjusting and mitigating the impact of climate change, protecting resources and environment, promoting carbon sequestration and emission reduction, and increasing farmers' income.

6.1.1 *China's Grain Demand Puts Forward New Requirements for Future Agricultural Production*

The results of *China's Agricultural Outlook Report* (2020–2029) showed that new achievements have been made in the high-quality development of China's agriculture. The report predicted that in 2020 the key goals in the field of "agriculture, rural areas and farmers" will be achieved as scheduled, the short board of agricultural and rural development will be constituted, and new achievements will be accomplished in the high-quality development of agriculture. Grain, pig, and other important agricultural products can achieve stable production and supply, and the effective supply guarantee capacity and supply quality of main agricultural products will be further improved. The agricultural planting structure will continue to be optimized, the variety structure of rice and wheat will be continuously optimized, the planting area will be appropriately reduced, the maize production capacity in the dominant production areas will be consolidated and improved, the planting area of maize in the "sickle bend" area will be slightly increased, and the planting area of rice, wheat, and maize will reach 95 million hectares. The soybean revitalization plan will continue with further implementation, with the area under soybean cultivation expected to increase by more than 133,000 hectares. The area of cotton cultivation will be stable at about 3.3 million hectares, rape planting area is expected to increase, and characteristic oil production is expected to increase to varying degrees. The consumption upgrading of urban and rural residents will accelerate, and the consumption trends of high-quality, green, personalized, and nutritious grains are more significant. The consumption of high-quality and special-purpose rice and wheat will grow rapidly, and the consumption demands on special food vegetable oils, high-quality characteristic fruits, and green and safe vegetables will be strong. The consumption of milk, poultry meat, and eggs will continue to grow, and the consumption structure of livestock products will be continuously optimized. Meanwhile, the consumption demand of processed agricultural products will increase steadily. The processed food consumption of fruits, potatoes, and bean products will increase by 3.9%, 1.8%, and 2% respectively. Due to the decrease in domestic consumption demand, the import of resource-intensive agricultural products such as cotton, edible vegetable oil, and sugar is expected to decline to varying degrees. The price of agricultural products is expected to rise in general, and the risk of high-level fluctuation of some varieties will further increase. The prices of rice and wheat are expected to fluctuate slightly, and there will be risks of international price transmission. The supply and demand of maize will be tight. It is predicted that the prices of agricultural products such as soybeans, cotton, sugar, edible vegetable oil, and other agricultural products will be determined by the international situation. The price of pork is likely to increase significantly, and this will keep the prices of livestock and poultry products high. The price of vegetables and fruits is expected to fluctuate greatly, and seasonal factors and unexpected events could still affect the price effectively. The price differences between regions, varieties, and quality will also be more significant.

In the next decade, the transformation and upgrading of agriculture will be significantly accelerated, the high-quality development of agriculture will achieve remarkable results, and the level of agricultural modernization will be steadily improved. Agriculture will be shifted from its present orientation on production rises to quality improvement. The production and supply of green, ecological, and safe agricultural products with high quality will increase significantly, and the national capacity of increasing production and ensuring the supply of grain and other major agricultural products will be greatly improved. With the decisive role of the market in allocating resources, the agricultural production structure guided by market demand is going to be continuously optimized, and the supply of market-oriented varieties and products meeting consumer demand will increase; the structure of ration varieties will be continuously optimized, and the area will be steadily decreased. The growth of unit yield will promote the steady yield growth of rice and wheat, with an average annual growth rate of less than 1%. The average annual yield growth rate of soybean and milk production of 3% will be achieved. The average annual yield growth rate of maize, fruit, and poultry is predicted to reach 2–3%. The total consumption of grain and other major agricultural products will continue to grow, the structure of food consumption will continue to upgrade, and the consumption of feed grain and industry will continue to increase. The increase of population will continue to increase the consumption of ration. In the next 10 years, the total domestic consumption of rice and wheat is expected to increase by 2.4% and 11.8%, respectively. The continuous expansion of animal husbandry production scale will promote the continuous growth of coarse grain feed consumption and soybean crushing consumption. The domestic total consumption of corn and soybean is expected to increase by 18.7% and 14.5%, respectively. The upgrading of food consumption structure will increase the consumption demand of animal products. The total domestic consumption of meat, eggs, dairy products, and aquatic products is expected to increase by 20.7%, 8.9%, 39.5%, and 9.8%, respectively.

The price formation mechanism of agricultural products is constantly improving, and the nominal and actual prices of agricultural products are on the rise. With the establishment and improvement of modern agricultural market system, the price of agricultural products will be mainly determined by the market. Under the irreversible conditions of the cost rise of labor, land, material input, and environmental protection, the price of agricultural products will rise integrally. With abundant grain ration supply, the marketization of rice and wheat prices will be more obvious. The supply of feed grain is tight, and the price of corn is expected to rise moderately. The land-intensive products such as soybean, cotton, sugar, and edible vegetable oil will be in short supply in China, the import dependence will become high, and the price linkage between domestic and foreign countries is going to be very obvious. The labor-intensive products such as vegetables, fruits, meat, eggs, dairy products, aquatic products, and so on will be driven by cost and market demand, and the prices will be generally on the rise. The new epidemic in 2020 has significantly affected the global food trade pattern, and China's food security needs to be further guaranteed.

6.1.2 *Challenges of China's Food Production with Climate Change*

Since the 1980s, global climate change has gradually attracted the attention of people from all walks of life. Since 1990, the Intergovernmental Panel on Climate Change (IPCC) has issued five consecutive global climate change assessment reports. A large number of research studies have shown that the global temperature rise caused by anthropogenic greenhouse gas emissions has had significant impacts on global climate change, and with the intensification of human activities, the trend of global warming will continue to intensify (Stocker *et al.*, 2014). According to a recent IPCC assessment report, the global average temperature has risen by 0.85°C in the past 100 years. It was predicted that the global average temperature would rise by 3.7–4.6°C (Stocker *et al.*, 2014). In the background of global warming, China's climate change also showed obvious characteristics, among which there was a general rise of temperature in the whole country. While precipitation showed a decrease in the north, unbalanced precipitations appear within regions and seasons (Qin, 2004, 2014; Li *et al.*, 2007; Ge *et al.*, 2014). With increasing global warming, the frequency and intensity of extreme climate events also showed an increasing trend (IPCC, 2012).

A large amount of research results showed that climate change has had significant impacts on all aspects of human life (Burke *et al.*, 2015; Challinor *et al.*, 2014; Lobell *et al.*, 2011b; Moore and Diaz, 2015; Wu *et al.*, 2014). The growth and development of crops and the formation of final yield in the agricultural ecosystem showed a certain degree of dependence on external environmental conditions (Cai *et al.*, 2003; Rosenzweig *et al.*, 2014). This high dependence on natural conditions determines that any degree of climate change will have some impact on agricultural production. Therefore, how to correctly evaluate the impacts of climate change on agricultural production is an important issue that government departments and scholars from all walks of life pay attention to, and it is also an important task in agricultural production at present. As a large agricultural country, China's agricultural production plays an important role in maintaining the national economy and people's livelihood. With the trend of global warming, the impact of climate change on China's agricultural production has gradually emerged (Xiong *et al.*, 2009a, 2010; Piao *et al.*, 2010). The impact of climate change on agricultural production would be further intensified with global warming in the future, and the yield of main food crops in tropical and temperate zones will decline significantly if adaptive measures are not undertaken (Challinor *et al.*, 2014; Zhao *et al.*, 2016a, 2017a). In addition, it was also pointed out that climate resource fluctuation is the main contributing factor of the yield fluctuation of main grain crops at this stage (Ray *et al.*, 2015). Therefore, it was very important for government departments and farmers to take adaptive measures and formulate corresponding coping strategies to improve regional adaptability and ensure food security in the background of climate change (Olesen *et al.*, 2011; Yang *et al.*, 2015).

The first is the impact of climate change on crop phenology. Phenological period is an important indicator to reflect the process of crop growth and development, and the impact of climate change on it will reflect the final crop yield. Therefore, using phenological data to analyze the change of phenological period and its response to climate change has become a research hotspot. At present, studies on the response of phenological changes to climate change mainly focus on natural vegetation, which is less affected by human activities, while less research has been done on crops (Menzel, 2000; Siebert and Ewert, 2012; Ge *et al.*, 2015). Long time-series studies showed that temperature rise had an obvious effect on the change of vegetation phenology (Sacks and Kucharik, 2011; Atlin *et al.*, 2017), which was mainly manifested in that the increase of temperature will lead to shorter crop growth period and earlier phenological period (Thomas *et al.*, 2014; Van Bussel *et al.*, 2015; Zhang *et al.*, 2016b; Rezaei *et al.*, 2017). The change of crop phenological process caused by climate change will inevitably lead to the change of crop growth and development process, which will affect the synthesis and accumulation of organic matter and eventually lead to the change of yield (Bindi and Olesen, 2011; Tao *et al.*, 2013; Challinor *et al.*, 2016). With further research on the impact of climate change on agricultural production, scholars at home and abroad used historical meteorological data and crop phenological period data to analyze the quantitative relationship between the change of phenological period and the change of external environment. Using crop growth model, they quantitatively evaluated the changes of crop phenology and its effect on final crop yield under different scenarios. The second is the impact of climate change on crop yield. Making a quantitative and clear answer to the impact of climate change on food production is not only essential to ensure food security but also the main basis for formulating policies to deal with climate change (Yao, 2005). Climate change has had important impacts on agricultural production. In the studies of the impact of climate change on crop production, the major concern is about its impact on crop yield. Many scholars have carried out a number of evaluations from the field scale to the regional scale, which further confirmed the existence of climate change impact. The last is the impact of climate change on crop distribution. With climate change, crop distribution and cropping system have changed to some extent (Li *et al.*, 2015c; Olesen *et al.*, 2011; Zhang and Yang, 2016). Although the progress of management measures and the change of varieties have promoted the increase of crop yield, the possible impact of climate change on crop planting area and planting layout should not be ignored (Singh *et al.*, 2017a; Wang *et al.*, 2017a). Yang *et al.* (2010) found that with global warming, the planting north boundary of the main cropping systems in China has moved northward to varying degrees; Li *et al.* (2015c) pointed out that the temperature rise had significant impacts on the migration of rice-planting center of gravity and has obvious synchronization with the northward shift of the planting center. Using the maximum entropy principle of species distribution, Liu *et al.* (2015) a found that the rice suitability regions in China showed an increasing trend. These values were consistent with the actual observation results. In the analysis results of rice suitability regions, climate change was the main factor affecting the spatial and temporal distribution of rice-planting areas. When analyzing the potential suitable planting areas of rice in China in the future, it was found that the planting areas tend to move to the northwest. Considering the social environment and other factors, it is very important to balance the matching of rice planting layout and climate resources. At present, domestic and foreign scholars have also analyzed the possible effects of climate change from the perspective of planting boundaries. Yang *et al.* (2015) pointed out that the northward movement of planting boundaries can increase the production capacity of China's grain crops by 2.2%. Other scholars have also obtained similar results in the study of wheat and rice in different regions (Singh *et al.*, 2017a; Wang *et al.*, 2017a).

6.1.3 *Contradiction Between Intensive Agriculture and Resource Environment Protection*

China's per capita cultivated land area is only 40% of the world's evaluation standard; the per capita cultivated land area of 666 China's counties is lower than the 533 m² warning line formulated by FAO, and 463 counties are lower than the 333 m² danger line. In the recent decade, the average annual decrease of cultivated land area is still more than 333,500 hectares. Agricultural intensification is an important feature of China's agriculture and also is an important driving factor for the continuous improvement of agricultural output in China. With the rapid development of marketization, industrialization, and urbanization, as well as the continuous improvement of the degree of opening up at home and abroad, it is getting harder to achieve high yield by simple high investment, which not only weakens competitiveness in the international market but also cannot be supported by local governments and farmers. In recent years, China's food security, as a major political task, has realized "15 years of continuous growth" due to forced implementation, but the cost is also quite huge. Under the traditional agricultural-intensive mode, the sustainable increase of grain yield not only entails an increasing amount of resources but also increases the pressure of ecological environment, which is difficult to sustain. Therefore, it is necessary to build a new agricultural intensive mode.

The increase of irrigation area has made great contribution to support the increase in grain production in China. The effective irrigation area in China has grown from 44.888 million ha in 1980 to 63.333 million ha in 2012, increasing by 41.1%, with an annual growth of 1.28%. However, China is one of the countries with serious water shortage in the world. At present, the gap has exceeded 50 billion m³, of which agricultural irrigation water shortage is more than 30 billion m³ and water shortage has brought increasing effects to agricultural development and food production prominently. Only 17% of the food production in the United States needs irrigation. Even if it runs out of irrigation water, it could still manage. Even world's food supply will meet the challenges of water shortage and will not have disastrous consequences. But 70% of China's food is produced by irrigation, and if it is exhausted, the consequences would be disastrous.

Fertilizer input has been increasing rapidly in the past 50 years and has become the main basis of agricultural growth. In 1980, only 12.69

million tons of chemical fertilizer was applied in China, reaching 58.84 million tons in 2012, an increase of 4.6 times and an average annual growth of 14.4%. However, the utilization rate of fertilizer was very low. The average apparent utilization rate of nitrogen fertilizer in China was less than 30%, the cumulative utilization rate was less than 50%, and the loss rate was as high as more than 50%. Due to the excessive use of chemical fertilizers, ecological and environmental problems and quality problems of agricultural products have begun to appear.

With excessive use of farmland, the quality of cultivated land shouldn't be ignored. There are some problems in the main grain-producing areas, such as shallow plough layer, increased bulk density, and decreased nutrient efficiency. Moreover, the problems of cultivated land's ecological quality caused by unreasonable fertilization, cultivation, and plant protection have become increasingly prominent. The organic matter content of black soil in Northeast China decreased from 5–7% in the past to 3–5% at present. The average thickness of plough layer is only about 20 cm, which is 5–10 cm lower than that of black soil area in the United States. The degree of soil acidification in South China continues to increase. The pH value of some red soils decreased from 6.5 30 years ago to about 5.6 now, which not only reduced crop yield but also aggravated soil heavy metal pollution. Due to the long-term excessive application of herbicides and other pesticide preparations, the "toxic soil" phenomenon even occurred in some areas, which brings more serious ecological security problems.

The tendency of agricultural non-point source pollution caused by traditional agricultural intensification is aggravating. According to some surveys, the average inputs of pure nitrogen and phosphorus are 30–40 kg 667 m^{-2} yr^{-1} and 20–30 kg 667 m^{-2} yr^{-1}, respectively. The actual utilization rate is less than 40%. The annual pesticide dosage is more than 200,000 tons (converted into pure), but only 20–30% of them is effective, and the remaining 70–80% of the pesticides enters water and soil, which will become a serious pollution source. At present, there are nearly 13.33 million ha of arable land polluted by pesticides, of which more than 50 million mu are moderately polluted.

In addition, from the comparison of the chemical fertilizer use in grain production between China and the United States, the growth of fertilizer consumption in the United States is no longer obvious after the 1980s, while the use of chemical fertilizer in China is still shows a significant growth trend. Moreover, from the aspect of the apparent efficiency of fertilizer input (grain yield per unit fertilizer input), the fertilizer efficiency of China is low. In the past 30 years, the fertilizer efficiency of the United States is basically stable and even follows an upward trend. This indicates that after China's grain yield reaches 600 million tons, we can focus on controlling the excessive growth of input and improving resource efficiency.

6.2 CSA Can Help China's Agricultural Transformation and Development

6.2.1 *Adaptation to Climate Change Is the Key to Stable and Increase of Grain Yield*

At present, the trend of global warming is increasingly recognized in the academic community, and climate change has become a focus of attention. Climate change has brought severe challenges to agricultural production. Studies showed that global wheat and maize yields have been reduced by 5.5% and 3.8%, respectively, due to climate change, which largely offsets the increase in production caused by favorable factors such as technological progress (Lobell *et al.*, 2011a). Ray *et al.* (2015) pointed out that about one-third of the global food yield change was caused by the change of meteorological factors. Some scholars predicted that global warming in the second half of this century will lead to a stronger reduction in tropical crops than temperate crops, but even a small increase in global warming will significantly reduce the yield of temperate crops (Challinor *et al.*, 2014). For China's agricultural production, no matter the boundary shift of the planting system, the exertion of crop light and temperature production potential, or the frequency of natural disasters and the risk of diseases and insect pests were affected by climate change to varying degrees (Yang *et al.*, 2010; Liu *et al.*, 2012, 2013). The United Nations Intergovernmental Panel on Climate Change (IPCC) pointed out that, in general, rice and maize production in China would show different degrees of reduction with temperature rise in future, but the effect and degree of crop yield affected in different regions were not consistent (IPCC, 2014). Therefore, coping with and mitigating climate change has become an important direction of agricultural research. On the other hand, according to the latest forecast report of the United

Nations (2019), by the middle of this century, the world population would reach 9.7 billion, which undoubtedly increased the demand for food in the future. In increasingly severe food demand situation and deteriorating ecological environment, how to find a way out in the future requires a comprehensive consideration of agriculture and even the whole human behavior (Foley *et al.*, 2011; Tilman *et al.*, 2011; Tilman and Clark, 2014).

Agricultural production is significantly affected by global climate change, and its risk is increasing under uncertain climate scenarios in the future. In addition to the threat of uncertain climate and environment in the future, the continuous growth of food demand and the slowdown of crop yield growth rate also seriously threaten human survival (Foley *et al.*, 2011; Tilman *et al.*, 2011; Ray *et al.*, 2012). Agricultural production and land use are important emission sources of greenhouse gas, which, to a certain extent, increase greenhouse gas emissions and exacerbate global warming. According to the fourth assessment report of IPCC (2007), agricultural activities account for 13.5% of all greenhouse gas emissions caused by human activities. It was estimated that about 60% of CH_4 and 61% of N_2O in major greenhouse gas emissions came from agricultural production (IPCC, 2013). For the planting industry, CH_4 and N_2O emissions mainly came from direct emissions of soil and indirect emissions of agricultural production materials (fertilizer, diesel, seeds, and other inputs) (Lal, 2004a; Smith *et al.*, 2008; van Kessel *et al.*, 2013; Xue *et al.*, 2014). The warming potential of CH_4 and N_2O is 25 and 298 times higher than that of CO_2 on a hundred-year scale (IPCC, 2006, 2007), and the increase of CH_4 and N_2O emissions has a significant impact on climate change. Unreasonable agricultural management measures not only affected the growth and development of crops but also accelerate the loss and emission of soil carbon (Lal, 2004a, 2007). Due to land reclamation and long-term tillage disturbance, soil organic carbon in temperate zone and tropical zone has been lost by nearly 60% and 75%, respectively (Lal, 2004b, 2010). It was predicted that by the middle of this century, the food demand of human beings will be twice that of the present, and agricultural production will face great pressure (Foley *et al.*, 2011; Ray *et al.*, 2012). Therefore, under the trend of global warming, how to improve agricultural production further in order to cope with climate change and ensure food supply better, as well as realize the sustainable development of agriculture is an important concern across the world (Lal, 2004b, 2007, 2015b).

6.2.2 *Carbon Sequestration and Emission Reduction Is the Core Content of Agricultural Sustainable Development*

Farmland is not only a source of greenhouse gas but also an important carbon pool. It was estimated that soil carbon pool, as the third largest carbon pool in the world, had 1,220–1,550 Pg (Peta, 1015g) of organic carbon and 695–748 Pg of inorganic carbon in 0–1 m soil layer, while 2,376–2,450 Pg organic carbon in 0–2 m soil layer (Lal *et al.*, 1995; Batjes, 1996). According to Lal (2004), the carbon sequestration potential of 0.4–1.2 Pg farmland soil could be realized by using reasonable farmland carbon pool management strategies. This was about 15% of the total annual anthropogenic carbon emissions, which could effectively alleviate global warming. Although there were still some controversies (Schlesinger *et al.*, 2019), the mitigation of global warming by promoting soil carbon sequestration has gradually been regarded as an important and effective strategy, especially after the "four thousandth" carbon sequestration project was proposed at the Paris global climate change conference (COP21) (Lal, 2016; Minasny *et al.*, 2017; Soussana *et al.*, 2019). At the same time, soil organic carbon content was an important indicator of soil quality, which was of great significance for maintaining farmland soil productivity (Reeves, 1997; Sá and Lal, 2009; Brandão *et al.*, 2011). Increasing soil carbon sequestration to mitigate climate change has also attracted increasing attention from the international community. The recommended farmland management measures (RMP) and integrated nutrient management measures (INM) could effectively increase soil carbon sequestration, improve soil quality, and achieve the goals of mitigating climate change, ensuring crop productivity and maintaining food supply (Lal, 2004b). Studies showed that compared with traditional farmland management measures, no-tillage, straw returning, optimized fertilization measures, and organic fertilizer application could effectively improve soil carbon sequestration capacity (Xue *et al.*, 2013; Wei *et al.*, 2013; Zhang *et al.*, 2014; Zhao *et al.*, 2015b) and reduce farmland greenhouse gas emissions (West and Marland, 2002; Ussiri *et al.*, 2009; Yang *et al.*, 2010; Zhao *et al.*, 2016). At the same time, it could also effectively improve crop production efficiency and increase yield (Ding *et al.*, 2014; Liu *et al.*, 2014b; Zhao *et al.*, 2017; Bai *et al.*, 2019). Lal (2004b) found that the increase of soil carbon sequestration capacity is helpful in increasing crop yield. For peer ton increase of soil carbon pool, maize yield increases by 10–20 kg ha^{-1}, while wheat yield increased by 20–40 kg ha^{-1}. Therefore, reasonable integrated farmland management measures can improve soil carbon sequestration capacity, reduce greenhouse gas emissions, as well as keep and even promote crop production, which is of great significance to alleviate global warming and ensure food security.

As the largest developing country in the world, China is facing severe population pressure and development pressure. As one of the most populous countries in the world, China feeds 22% of the world's population with less than 7% of the world's arable land. With the deepening of China's reform and opening up, as well as economic development, higher requirements are put forward for the structure, product supply, and production efficiency of agricultural production. Increasing crop yield per unit area is an important way to increase total grain yield. According to the statistical data from FAO, China's grain output in 2018 increased by nearly 40% compared with that in 2000, and the unit yield was nearly twice as the world average level (FAO, 2019). However, the high output of China's agricultural production is based on high input. A large amount of agricultural production inputs, especially the large use of chemical fertilizers, not only improved crop yield but also brought a series of ecological and environmental problems (Schlesinger, 2010; Chen *et al.*, 2014; Li, 2015). In the 1990s, the N_2O emission from China was 32.3 Gg C yr^{-1}, which was nearly increased from 22 Gg C yr^{-1} compared with the 1950s (Zou *et al.*, 2009). As a result, acid rain and water eutrophication were becoming a serious problem, threatening human health (Zhang *et al.*, 2013a). In addition, with the further development of agriculture, it is increasingly important to improve the use efficiency of agricultural inputs, especially to reduce the input of rural labor force, develop the "light simplification" agricultural cultivation management measures, and the requirements for efficient and sustainable intensive agriculture. In recent years, FAO (2013) proposed to develop CSA to cope with climate change. CSA is an important strategy to achieve "three wins" in food security, climate change adaptation, and mitigation. It needs the cooperation of technology innovation, policy management, and science and technology and financial investment to achieve the goal of improving crop productivity and solving food security; at the same time, the human role in promoting farmland system adaptation and mitigating climate change should be brought into play. Reducing greenhouse gas emissions from agricultural production and increasing soil carbon sequestration are effective means to achieve this goal. Therefore, improving farmland management measures and developing CSA are of great significance for reducing farmland carbon emissions, coping with and mitigating climate warming, ensuring food supply, and sustainable agricultural development in China. Meanwhile, optimizing management measures, reducing greenhouse gas emissions and increasing carbon sequestration in farmland soil, and realizing agricultural low-carbon production are also important manifestations of China's active international responsibility.

6.2.3 CSA Is an Important Solution to Achieve Green Agriculture

In recent years, the government has continuously issued announcements to promote the green transformation of agriculture. In 2015, the Ministry of Agriculture and Rural Areas, the National Development and Reform Commission, and other departments issued the *National Agricultural Sustainable Development Plan* (2015–2030). In 2017, the General Office of the CPC Central Committee and General Office of the State Council issued *The Opinions on Innovating Institutional Mechanisms to Promote Green Agricultural Development*. In 2018, the Ministry of Agriculture and Village issued *The Technical Guidelines for Agricultural Green Development* (2018–2030) to vigorously promote China's agricultural green-color development.

Green development is the inherent requirement of modern agricultural development and an important part of the construction of ecological civilization. In recent years, China's grain harvest year after year has been phenomenal and the supply of agricultural products is abundant. Agricultural development is constantly stepping to a new level. However, due to excessive use of chemical fertilizers and pesticides, the low utilization rate of livestock manure, crop straw, and agricultural film, and the excessive intensity of fishing, agricultural development is facing increasing resource pressure and ecological environment is turning into "red light". China's agriculture has come to a new stage of accelerating transformation and upgrading and realizing green development. The implementation of green development is conducive to promoting the comprehensive treatment and resource utilization of agricultural production wastes, slow down the excessive utilization intensity of agricultural resources, and the aggravation of non-point source pollution and promote the sustainable development of China's agriculture. General Secretary Xi Jinping pointed out that promoting the structural reform of the supply side of agriculture should give prominence to the supply of green high-quality agricultural products. At present, the supply of agricultural products in China is more than that of high-quality brands, which is not suitable for the rapid upgrading of consumption structure of urban and rural residents. To promote the green development of agriculture is to develop standardized and brand agriculture, provide more high-quality, secure and characteristic agricultural products, and also promote the supply of agricultural products from mainly meeting the demand of "quantity" to paying more attention to "quality". The implementation of green development is conducive to changing the traditional mode of production, reducing the excessive use of inputs such as fertilizer, optimizing the environment of producing areas of agricultural products, effectively improving the quality of products, and ensuring the supply of high-quality green agricultural products from source. Agriculture and environment are the most compatible. The beautiful environment of new countryside cannot do without the green development of agriculture. In recent years, with the rapid development of agricultural production, agricultural non-point source pollution is becoming more and more serious, especially livestock and poultry breeding waste pollution, which has a great impact on farmers' life and rural environment. General Secretary Xi Jinping emphasized that speeding up the disposal and recycling of livestock and poultry waste, which is related to the production and living environment of more than 600 million rural residents, is a good thing to benefit the country and people. The implementation of green development is conducive to reducing agricultural production waste emissions, beautifying the rural living environment, promoting the construction of new rural areas, realizing the harmonious development of man and nature, and the coordination and win–win situation of agricultural production and ecological environment.

The green development of agriculture is a profound revolution in the concept of agricultural development in China. Agricultural development should be changed from mainly meeting the demand of "quantity" to paying more attention to "quality". We should use limited resources to increase the supply of high-quality and safe agricultural products, reduce the excessive use of agricultural resources, slow down the trend of agricultural non-point source pollution, and make ecological and environmental protection a distinctive symbol of modern agriculture. At present, the five actions of green development in China's agriculture mainly include the following:

(1) Resource utilization of livestock manure. We should speed up the construction of a new pattern of sustainable development with the combination of planting and breeding and the circulation of agriculture and animal husbandry. We should carry out the utilization of livestock manure in animal husbandry-based counties, implement the integrated project of planting and breeding, promote the technical mode of livestock and poultry manure resource utilization, improve the treatment capacity of livestock and poultry manure, and strive to solve the problem of large-scale livestock and poultry farm manure treatment and recycling.

(2) Use organic fertilizer made from fruits, vegetables, and tea instead of chemical fertilizers. In order to develop ecological circular agriculture and improve the quality and efficiency of fruit and vegetable tea, we should vigorously promote the technology of replacing chemical fertilizer with organic fertilizer, accelerate the resource utilization of livestock and poultry breeding wastes and crop straw, and achieve cost-effective, quality, and efficiency improvement. Treat and utilize the crop straw.

(3) We should vigorously promote the utilization of straw as fertilizers, feeds, fuels, raw and basic materials, strengthen the research and development of new technology and new equipment, accelerate the establishment of industrialization utilization mechanism, and continuously improve the comprehensive utilization level of straw. We should promote the introduction of subsidy policies for straw returning, collection, storage, and transportation and also processing and utilization, in order to build a market-oriented operation mechanism.

(4) Recycle agricultural films. Taking the application of thickened plastic film, mechanized picking up, specialized recovery, and resource utilization as the main direction of attack, it is implemented in a continuous manner, promoted by the whole county and managed comprehensively. We should promote the use of thickened mulch film in an all-round way, promote the reduction of replacement, promote the establishment of recycling mechanisms in various ways, such as replacing the old with the new, handing in business entities, specialized organization recycling, and processing enterprise recycling, and improve the monitoring network of farmland residual plastic film pollution.

(5) Protect the aquatic organisms in key waters. We should adhere to the principle of ecological priority and green development, reducing ships and transfer production, gradually promoting the comprehensive prohibition of fishing in the Yangtze River Basin, taking the lead in implementing the ban on fishing in aquatic biological protection zones, and then restore the ecological environment of coastal fisheries along the river. We should carry out the demonstration of healthy aquaculture, strengthen the total management of marine fishery resources, and strengthen the protection of aquatic habitats.

Climate-smart agricultural production is a new concept of agricultural development. It can not only slow down climate change but also adapt to the adverse effects of climate change. Its core goal is to sequester carbon and reduce emissions, stabilize grain production, and increase income. In addition, improving soil health and enhancing soil's buffering capacity against climate change are important tasks of climate-smart agricultural production, and they are the core of achieving modern green agricultural development.

6.3 Relevant Basis of China's Adaptation and Mitigation of Climate Change

6.3.1 *China's Theories and Policies in Response to Climate Change*

In recent years, many studies on CSA have been done. Because the concept is understood differently, the focus of research has also differed in research. First, different scholars have different understanding of the definition and connotation of CSA. FAO defines it as the agricultural production and development mode aimed at improving agricultural efficiency, enhancing adaptability, reducing greenhouse gas emissions, and achieving national food security with higher goals, while the World Bank defines it as building a food system that can meet the growing demand and maintain profitability and sustainable development against climate change. Summarizing the consistent understanding, the CSA means the form and mode of agricultural development with high levels, high standards, and scientific response to climate change, which can ensure food production and improve the function of farmland ecosystem. It is an important reference for guiding the reform and adjustment of the agricultural system. It is of great significance to ensure food security, effectively support agricultural sustainable development, and adjust agricultural structure according to local conditions.

For the research focus of CSA, some scholars mainly focus on the change of crop yield. For example, Bai *et al.* (2016) used the APSIM model to simulate the change of wheat yield under different management scenarios from 1981 to 2010 and found that the climate change intelligent management model based on optimal planting density and sowing date, combined with good water and fertilizer management and plant protection measures, could significantly increase crop yield and stability. More scholars were concerned about the effects of CSA on carbon sequestration and emission reduction. At the 2015 Paris Climate Change Conference, France put forward a motion called "four thousandth plan", which focused on food security and climate. The plan linked soil carbon increase with sustainable agricultural production. It pointed out that increasing soil carbon pool through good farmland management measures could not only offset fossil fuel combustion emissions but also guaranteed production through increasing organic matter and improving soil quality (Lal, 2016; Cheng and Pan, 2016; Minasny *et al.*, 2017). With the help of meta-analysis method, Bai *et al.* (2019) evaluated the soil carbon sequestration effect of "climate change smart agriculture" management measures such as conservation tillage, mulching crops, and biochar application and found that biochar application was the most effective soil carbon sequestration measure, followed by covering crops and conservation tillage. In addition, studies have also focused on the important role of water management in future climate change response. For example, Brouziyne *et al.* (2018) used the Swat hydrological model, which was developed by the United States Department of Agriculture, to simulate the impact of different climate change scenarios on water yield and crop water productivity (yield per unit of actual evapotranspiration) of a watershed in northwest Morocco. At the same time, different optimal management scenarios were set up to explore the optimal intelligent management mode of climate change. It was found that future climate change would reduce water output by 26% and crop water productivity by 45%. The combination of no-till and early sowing date could effectively improve crop water use efficiency under future climate change scenarios. Generally, the current research on CSA mainly focuses on the use effect and advantages of different farming technologies, and the implementation approaches of it.

For a long time, the primary purpose of China's agriculture is to ensure food security. China's contribution to world food security is obvious for all. But at the same time, it is not difficult to see that the development of China's agriculture is to, a large extent, at the expense of resources and ecological environment. Meanwhile, development also brings about the shortage of natural resources, the aggravation of ecological environment problems, and low efficiency of agricultural production. During the 12th Five Year Plan period, China has strengthened disaster resistance and mitigation of agricultural production, increased grassland protection and recovery, promoted the transformation and upgradation of rural biogas, carried out comprehensive utilization of straw, promoted coal/fuel-saving stove, developed rural solar energy and microwater power, implemented zero growth action of fertilizer and pesticide use, implemented conservation tillage, and carried out fishery energy conservation. At present, various departments in China attach great importance to the high carbon emission and excessive greenhouse gas emission caused by agricultural production. In recent years, a series of technical research, project demonstration, and promotion and response measures have been carried out, focusing on agricultural carbon sequestration and emission reduction, coping with the impact of climate change and sustainable improvement of agricultural production efficiency, and they have achieved preliminary results.

As a big agricultural country, greenhouse gas emissions and food security in agricultural production have attracted the attention of governments at all levels in China. In recent years, a series of policy and technology research, demonstration, and extension works have been carried out around carbon sequestration and emission reduction in agriculture and response to climate change, and preliminary results have been achieved. In terms of policy, the National Development and Reform Commission, together with the Ministry of Finance, the Ministry of Agriculture, and other eight ministries and commissions, jointly issued the *National Strategy for Adapting to Climate Change* in 2013. The strategy clearly defines the direction of efforts for agriculture to adapt to climate change: the first is to strengthen agricultural monitoring and early warning and disaster prevention and mitigation measures; the second is to improve the adaptability of planting industry; the third is to guide the rational development of livestock and poultry and aquaculture industry; the fourth is to strengthen agricultural development security. In 2014, the National Development and Reform Commission issued the *National Climate*

Change Plan (2014–2020), which provided ideas for agricultural control of greenhouse gas emissions and adaptation to climate change. In 2016, the State Council issued the *13th Five-year Plan for Greenhouse Gas Emission Control*, which clearly proposed to develop low-carbon agriculture and increase ecosystem carbon sink. In addition, in terms of policy and system innovation, China has begun to carry out pilot demonstration work on technology subsidy, incentive mechanism, and policy guidance.

6.3.2 *China's Carbon Sequestration and Emission Reduction Related Technologies*

6.3.2.1 *Technologies for Conservation Agriculture*

Soil tillage is an important part of farmland management. Conventional tillage, such as ploughing, has a long history both in China and the world. Tillage could change soil structure and microenvironment and affect soil physical, chemical, and biological indicators, such as increasing soil porosity and aeration, reducing soil compaction, increasing infiltration, controlling weeds, and so on. Thus, it could affect the growth and development of crops, which plays a very important role in the development of world agriculture (Benites and Ofori, 1993; Lal and Kimble, 1997; Derpsch, 2004; Lal, 2009). However, frequent tillage also brought about such problems as large loss of organic matter and serious soil erosion by wind and water, which was not conducive to the sustainable use of farmland soil and even causes disastrous consequences. In the 1930s, the continuous high intensity of ploughing farmland soil with wall plow in the United States caused soil drought and wind erosion intensified in the Great Plains. A worldwide shock of "Black Storm" blew away millions of tons of surface soil and seriously degraded the fertility of 40 million hectares of farmland, resulting in huge losses (Logan *et al.*, 1991; Lal *et al.*, 2007). After reflection, the United States begun to study and promote the application of no-till technology. Later, the US Soil Protection Agency proposed a modern conservation tillage technology system based on no-till and promoted it worldwide. Zhang *et al.* (2014) systematically summarized the application and development process of conservation tillage in China and put forward conservation agriculture and its four important components, including less no-till, permanent mulching, diversity complex planting system, and integrated nutrient management system. Lal (2015a) summarized the research and application of global conservation tillage since the 1960s and pointed out that conservation tillage is a technology system with no-till as the main body. The potential application risks or defects should be avoided by using appropriate measures (straw mulching, mulching crops, crop rotation, integrated nutrient management, etc.) according to the actual local conditions.

With increasing attention to climate change, in order to achieve the goal of sustainable development, more attention is paid to the sustainable intensification of agricultural development, the ability to resist climate change, ecological efficiency, and the relationship between agricultural measures and soil management (United Nations, 2014). Conservation tillage with no-till as the core has been widely used as an important part of sustainable agriculture in the world (Baker *et al.*, 2007; Derpsch, 1998, 2004; Friedrich *et al.*, 2012; Lal, 2013; Zhang *et al.*, 2014). The advantages of conservation tillage mainly include promoting soil and water conservation, improving soil structure, improving soil quality, maintaining crop production, and strengthening farmland carbon sequestration and emission reduction (West and Post, 2002; Zhang *et al.*, 2014; Lal, 2015a, 2015b). Some scholars have estimated that 122–215 million ha of land around the world have adopted protective agricultural measures, accounting for about 9–15% of the global cultivated land area. At the same time, nearly 53.3–113 billion ha of land (about 38–81% of the global cultivated land) hold the potential to use protective agricultural technology measures (Prestele *et al.*, 2018). In recent years, the research on crop yield, greenhouse gas emissions, climate change, and ecological services of conservation tillage have become the hot topic in the world (Lal, 2013; Palm *et al.*, 2014; Powlson *et al.*, 2014; Pittelkow *et al.*, 2015a).

China is a big agricultural country with a long history of intensive farming. However, long-term tillage degenerates the quality of farmland soil, which restricts the sustainable development of agricultural production. The research on conservation tillage in China began in the 1970s, and the conservation tillage with no-till as the core has been popularized and applied in China with its characteristics of soil conservation, fertilizer cultivation, water saving, and labor saving. At present, the extension area of conservation tillage technology has exceeded 7.33 million ha. Reviewing the development history of conservation tillage technology in China, China has a long history of water and soil conservation in technical agricultural measures, but the exploration of mechanized conservation tillage began in the early 1980s. At that time, drawing on the experience of foreign countries, we carried out the test and research work of no-till, subsoiling, mulching, and other single technologies. However, due to China's long-term advocacy of Chinese traditional intensive farming, although the experimental research of no-till and mulching technology has achieved satisfactory production results, it has not formed a large application environment. Since 1992, the Australian conservation tillage technology has been introduced and used as a reference in Shanxi Province. And since 2001, Gansu Agricultural University, Lanzhou University, and the University of Adelaide have started active explorations in the Loess Plateau region of western Gansu.

Based on the summary of the existing studies on conservation tillage in China, it was found that at present, a lot of attention had been paid to the effects of conservation tillage on the soil's physical and chemical properties (such as soil bulk density, soil moisture, and temperature) (Zhang *et al.*, 2005; Du *et al.*, 2010), crop yield (Xie *et al.*, 2008), soil carbon sequestration (Liang *et al.*, 2011; Zhang *et al.*, 2014), soil quality (Huang *et al.*, 2010), greenhouse gas emissions (Zhang *et al.*, 2015; Zhao *et al.*, 2016), and farmland ecosystem carbon footprint and ecological service value (Xue *et al.*, 2014; Zhang *et al.*, 2016). Zhang *et al.* (2014) summarized the challenges and future opportunities encountered in the development of conservation tillage (agriculture) in China and pointed out that diversified planting systems, the phenomenon of reduced and no-tillage and yield reduction, lack of planting machinery suitable for small-holder operations, and people's lack of understanding of conservation tillage (agriculture) are the main challenges facing the development of conservation tillage (agriculture) in our country. At the same time, they also pointed out that our country's agriculture is currently facing a high degree of intensification, declining soil fertility, large amount of crop straw, and agricultural economic benefits. Due to the problems of low labor force and shortage of rural labor force, conservation tillage has a good development prospect in China due to its many unique functions.

At present, the effect of conservation tillage on crop yield is controversial, and there are obvious differences among different research results. The results of global scale meta-analysis showed that compared with traditional tillage, no-till straw returning reduced crop yield, but combined with straw returning and crop rotation, crop yield increased (Pittelkow *et al.*, 2015a). In addition, long-term use of no-till or reduced tillage in arid areas improved crop yield compared with short-term no-till (Van Kessel *et al.*, 2013). Compared with no straw returning, straw returning has significantly improved the yield of rice and dry farming crops (Liu *et al.*, 2014a). Xie *et al.* (2008) summarized the research on the impact of conservation tillage measures on crop yield in China and found that compared with the traditional tillage, the crop yield increased by 12.5% on average, out of which the rice yield increased by 15.9%; on the other hand. Nearly 10% of the studies showed that the crop yield showed a downward trend, especially in North China, where the yield reduction rate of wheat was as high as 18.5%. This showed that the application of conservation tillage measures in different regions and different crop types had different effects on crop yield (Xie *et al.*, 2008; Zhao *et al.*, 2017). There was a consistent understanding of the positive effect of straw returning on crop yield. The results of meta-analysis at the national scale showed that straw returning could effectively improve crop yield in China (Huang *et al.*, 2013; Zhao *et al.*, 2017).

Soil is an important carbon pool. Enhancing soil carbon sequestration capacity plays an important role in coping with and mitigating climate change and maintaining food security. Farmland management measures, such as soil tillage (West and Post, 2002; Ussiri and Lal, 2009; Dalal *et al.*, 2011; Zhang *et al.*, 2014), straw management (Lu *et al.*, 2009; Ding *et al.*, 2014; Liu *et al.*, 2014a), chemical fertilizer application (Lu *et al.*, 2009; Ding *et al.*, 2014), organic fertilizer input (Ding *et al.*, 2014; Maillard and Angers, 2014), and soil water management measures (Abid and Lal, 2008) will affect soil organic carbon pool. A large number of studies showed that compared with conventional tillage, conservation tillage measures (such as no-till straw returning) significantly improve the fixation capacity of soil organic carbon, especially the surface soil organic carbon fixation (West and Post, 2002; Dalal *et al.*, 2011; Xue *et al.*, 2013; Zhang *et al.*, 2014). The increase of soil carbon pool is conducive to further improvement of soil quality and soil structure, thus changing the farmland soil microenvironment. Therefore, conservation tillage measures are conducive to promoting crop growth and development (Delgado *et al.*, 2013; Zhang *et al.*, 2014). Some results showed that the improvement of soil carbon sequestration capacity is conducive to the increase of crop yield, especially in the soil with serious quality degradation, and the increase of soil organic carbon content has obvious effect on crop yield (Lal, 2004b; Sá *et al.*, 2014).

The results of no-till straw returning and other conservation tillage measures on the direct emission of CH_4 and N_2O in farmland soil varied and were significantly affected by climate and soil factors of the study site. Some results showed that no-till straw returning could change soil indicators (such as pH value, water content, organic carbon and its components, nitrogen status, bulk density, porosity), organic matter degradation, and microbial utilization (Six *et al.*, 2002; West and Post, 2002; He *et al.*, 2011; Zhang *et al.*, 2013c; Liu *et al.*, 2014a; Mangalassery *et al.*, 2014). In this way, CH_4 and N_2O emissions are affected differently (Stevens *et al.*, 1998; Godde and Conrad, 2000; Ussiri *et al.*, 2009; Ruan and Robertson, 2013; Yao *et al.*, 2013). In order to reveal the changes and main influencing factors of greenhouse gas emissions caused by an agricultural measure, the meta-analysis method has been widely used to evaluate greenhouse gas emissions from soil under conservation tillage (Chen *et al.*, 2013; Van Kessel *et al.*, 2013; Skinner *et al.*, 2014). Van Kessel *et al.* (2013), based on the results of meta-analysis on a global scale, pointed out that no-till or reduced tillage increased N_2O emissions but did not explain the variation of N_2O emissions at different research sites and the causes. There are a few reports on the impact of agricultural measures on soil CH_4 and N_2O emissions at the regional or national scale in China. In addition, meta regression method has been widely used to explore the relationship between greenhouse gas emissions and its influencing factors (Gattinger *et al.*, 2012; Skinner *et al.*, 2014). It is necessary to study the effects of conservation tillage on soil CH_4 and N_2O emissions and its influencing factors at the national scale with the help of meta-analysis tools. Some results showed that no-till, straw returning, and other conservation tillage measures could eliminate the adverse effects of increased greenhouse gas emissions through improving soil carbon sequestration capacity, which was manifested as net carbon sequestration (Zhang *et al.*, 2013c; Liu *et al.*, 2014a). In addition to the changes of soil carbon pool and direct greenhouse gas emissions, greenhouse gas emissions caused by the production and use of agricultural inputs also had a certain impact on the total carbon emissions of farmland ecosystem. It is generally believed that conservation tillage can significantly reduce indirect greenhouse gas emissions from farmland by reducing fertilizer input and mechanical use.

6.3.2.2 *Straw Returning Technology*

Crop straw is an important agricultural product. It is estimated that the annual crop straw yield was about 3.8 Gt (109 t) (Lal, 2005). According to Song Dali *et al.* (2018), China's crop straw output was about 720 million tons per year, which can effectively replace the use of some chemical fertilizers. At present, there are many ways to utilize crop straw, such as direct returning to field, animal feed, fuel, and raw materials of biomass energy (Lal, 2008; Sun *et al.*, 2016; Li *et al.*, 2017). Among these utilization methods, direct straw returning is taken as the most economical environmental protection measure. On the one hand, it has brought certain nutrient supplement to soil; on the other hand, it could significantly improve soil carbon fixation capacity, thus effectively improving soil quality and productivity, and plays an important role in maintaining food security and mitigating climate change (Lal, 2004; Li *et al.*, 2019). At the same time, as one of the core technologies of conservation agriculture, straw returning technology has been widely used in China and even in the world (Zhang *et al.*, 2014; Lal, 2015). However, due to the obvious differences in climate, soil types, and planting systems in different regions of China, the effects of straw returning must also have obvious differences, which greatly affects the popularization and application of straw returning technology. Therefore, it is important to explore the impact of straw returning technology on farmland soil quality, soil carbon sequestration capacity, greenhouse gas emissions, environmental impact, and crop production. And we should systematically analyze its changes in different regions, which is of great significance to clarify the "positive and negative" effects of straw returning and put forward the "negative" effect avoidance scheme. It can also provide certain theoretical support for the promotion of straw returning technology and the realization of sustainable development of agricultural production, which is of great significance for the development of "climate-smart" agriculture.

Straw mainly refers to all the residual by-products in the field after harvesting the main crop products and the by-products produced during the primary processing of the main products. According to different sources, they have been divided into field straw, processing by-product, field crop straw, and horticultural crop straw (Xie *et al.*, 2010). Generally, crop straw usually refers to field crop straw. This report focuses on the utilization of crop straw and the effect of returning it to the field. In terms of the total amount of resources, the amount of crop straw resources in China is large, and it is increasing every year. In the 1990s, China's crop straw output was about 60–680 million tons; in the first 10 years of the new century, China's crop straw output reached about 740 million tons; according to the estimation of scholars, China's crop straw output in 2016 was about 820 million tons (Xie Guanghui *et al.*, 2010; Li *et al.*, 2017; Yu and Yang, 2018). Within different crop types, rice, wheat, maize, and other crops are the main sources of crop straw, and the distribution is uneven in different regions (Song *et al.*, 2018). In terms of crop species, maize, rice, wheat, oil crops (rape, peanut, soybean), cotton, and potato accounted for 41.7%, 23%, 15.8%, 10.9%, 2%, and 1.8% of the total crop straw, respectively, in China. From a regional perspective, North China, Northeast China, middle and lower reaches of the Yangtze River, Southeast, Southwest and Northwest China contributed 26.4%, 19%, 26.2%, 2.2%, 11.7%, and 11.5% to the national straw production, respectively (Song *et al.*, 2018).

In the face of such a huge resource potential, how to make better use of this resource is also crucial. Chinese government encourages the resource utilization of crop straw and summarizes the five utilization methods of straw resources: energy (or fuel), fertilizer, feed, ingredients, and basic materials. Since the general office of the State Council issued the *Opinions on Accelerating the Comprehensive Utilization of Crop Straw* in 2008, the utilization level of straw resources has been greatly improved. At present, the comprehensive utilization rate of straw in China has exceeded 65% (Wang *et al.*, 2018). Specifically, according to estimation by scholars, the amount of straw consumed by different straw utilization methods, such as direct return to field, animal feed, supplementary fuel, biomass energy, industrial raw materials, waste, or incineration accounted for 14.1%, 27.1%, 32.3%, 1.1%, 8.6%, and 16.8% of the total straw resources, respectively (Bi *et al.*, 2019). Among various utilization methods, straw returning directly is the most simple, cheap, and efficient treatment method. In recent years, with the increase of national policy support and farmers' willingness to accept, the area of straw returning to the field is increasing. Straw returning is one of the main ways of crop straw fertilizer utilization, including direct return and indirect return. Direct returning includes direct ploughing, with mulching and smashing of straw; indirect returning includes fast decomposition returning, composting returning, over belly returning, dregs returning, and straw making into an organic compound fertilizer.

Reviewing the development of straw returning technology in China, it mainly experienced the following stages: at the end of 1970s, as domestic scholars began to pay attention to conservation tillage technology, no-till and straw returning as two core technologies were introduced, studied, promoted, and applied. After a period of research and application, towards the end of the 1990s, the Ministry of Agriculture (now the Ministry of Agriculture and Rural Areas) carried out research on straw returning technology, led by China Agricultural University, organized and carried out the national survey, research, and application of straw returning technology in China, and finally compiled the book *Mechanism and Technical Model of Straw Returning*. It laid a solid foundation for the in-depth research and application of straw returning technology in China. Since then, during the period of "Eleventh Five Year Plan" and "Twelfth Five Year Plan", with the development of conservation tillage and circular agriculture, the technical mode and mechanism of straw returning in different regions, crops, planting patterns, and soil and climate conditions in China were systematically and deeply studied. At present, according to the data released by the Agricultural Mechanization Department of the Ministry of Agriculture and Rural Areas, as of 2016, the mechanized straw returning area in China exceeded 480 million hectares, accounting for 28.8% of the total sown area (Agricultural Mechanization Department of the Ministry of Agriculture and Rural Areas, 2017).

Based on the summary of research on straw returning technology in China, it has been found that research mainly focused on fertilizer substitution, soil's physical and chemical properties, soil's carbon pool dynamics, greenhouse gas emissions, farmland sustainability, ecological service value, and ecological footprint. However, there were also many problems in the development of straw returning technology in China: too much straw was returned to the field; the quality and time of returning were not suitable; the straw crushing technology was not qualified; the selection and improvement speed of new agricultural machinery were too slow; the decomposition of straw was slow and the accumulation was too large; the unsuitable cultivation mode affected the seedling situation; the field management measures were not timely, which affected the crop population quality; the integration technology of agricultural machinery and agronomy didn't reach the standard; the policies and regulations of straw returning to field were incomplete; the willingness of farmers to accept was low, which restricted the wide application of the technology (Zhang *et al.*, 2014; Zhao *et al.*, 2017; Yu and Yang, 2018; Bi *et al.*, 2019). However, this also shows that straw returning technology in China still has a great potential and it has a good development prospect. In the future, with the deepening of research on protective agricultural technology and the increasing demand of green agriculture development, the combination of straw returning technology and other coordination measures has great significance for the construction of CSA.

References

Agricultural Mechanization Department of the Ministry of Agriculture and Rural Areas. 2017. The 13th Five-Year Plan for the Development of National Agricultural Mechanization. https://www.moa.gov.cn/nybgb/2017/dyiq/201712/t20171227_6129917.htm

Atlin G N, Cairns J E, Das B. 2017. Rapid breeding and varietal replacement are critical to adaptation of cropping systems in the developing world to climate change. *Global Food Security*, 12: 31–37.

Bai H, Tao F, Xiao D, *et al.* 2016. Attribution of yield change for rice-wheat rotation system in China to climate change, cultivars and agronomic management in the past three decades. *Climate Change*, 135(3–4):539–553.

Bai X, Huang Y, Ren W, *et al.* 2019. Responses of soil carbon sequestration to climate-smart agriculture practices: A meta-analysis. *Global Change Biology*, 25: 2591–2606. https://doi.org/10.1111/gcb.14658.

Baker J M, Ochsner T E, Venterea R T, *et al.* 2007. Tillage and soil carbon sequestration — What do we really know? *Agriculture Ecosystems & Environment*, 118 (1–4): 1–5.

Batjes N H. 1996. Total carbon and nitrogen in the soils of the world. *European Journal of Soil Science*, 47 (2): 151–163.

Bi Y Y, Gao C Y, Wang H Y, et al. 2019. A summarization of national regulations on the comprehensive utilization of crop straw and the management of burning ban and legislative suggestions. *China Agricultural Resources and Regional Planning*, 40(8):1–10.

Bindi M, Olesen J E. 2011. The responses of agriculture in Europe to climate change. *Regional Environmental Change*, 11, S151–S158.

Brandão M, Milài Canals L, Clift R. 2011. Soil organic carbon changes in the cultivation of energy crops: Implications for GHG balances and soil quality for use in LCA. *Biomass and Bioenergy*, 35 (6): 2323–2336.

Burke M, Hsiang S M, Miguel E. 2015. Global non-linear effect of temperature on economic production. *Nature*, 527: 235–239.

Cai Z C, Tsuruta H, Gao M, *et al.* 2003. Options for mitigating methane emission from a permanently flooded rice field. *Global Change Biology*, 9 (1): 37–45.

Challinor A J, Koehler A K, Ramirez-Villegas J *et al.* 2016. Current warming will reduce yields unless maize breeding and seed systems adapt immediately. *Nature Climate Change*, 6, 954–958.

Challinor A J, Watson J, Lobell D B, *et al.* 2014. A meta-analysis of crop yield under climate change and adaptation. *Nature Climate Change*, 4 (4): 287–291.

Chen D G, Peace K E. 2013. *Applied Meta-analysis with R*. Boca Raton, USA: CRC Press.

Chen H, Li X, Hu F, *et al.* 2013. Soil nitrous oxide emissions following crop residue addition: A meta-analysis. *Global Change Biology*, 19 (10): 2956–2964.

Chen H, Wang J, Huang J. 2014. Policy support, social capital, and farmers' adaptation to drought in China. *Global Environmental Change-Human and Policy Dimensions*, 24, 193–202.

Dalal R C, Allen D E, Wang W J, *et al.* 2011. Organic carbon and total nitrogen stocks in a Vertisol following 40 years of no-tillage, crop residue retention and nitrogen fertilisation. *Soil & Tillage Research*, 112 (2): 133–139.

Derpsch R. 1998. Historical review of no-tillage cultivation of crops. Proc. 1st JIRCAS Seminar on Soybean Research. *No-tillage Cultivation and Future Research Needs*, 13:1–18.

Derpsch R. 2004. History of crop production, with and without tillage. *Leading Edge*, 3:150–154.

Derpsch R, Franzluebbers A J, Duiker S W, *et al.* 2014. Why do we need to standardize no-tillage research? *Soil and Tillage Research*, 137: 16–22.

Ding X L, Yuan Y R, Liang Y, *et al.* 2014. Impact of long-term application of manure, crop residue, and mineral fertilizer on organic carbon pools and crop yields in a Mollisol. *Journal of Soils and Sediments*, 14 (5): 854–859.

Du Z, Ren T, Hu C. 2010. Tillage and residue removal effects on soil carbon and nitrogen storage in the North China Plain. *Soil Science Society of America Journal*, 74 (1): 196.

FAO. 2013. FAO Statistical Yearbook (2013) World Food and Agriculture Organization, Rome.

FAO. 2019. The state of food security and nutrition in the world. http://www.fao. org/state-of-food-security-nutrition/zh/.

Foley J A, Ramankutty N, Brauman K A *et al.* 2011. Solutions for a cultivated planet. *Nature*, 478: 337–342.

Gattinger A, Muller A, Haeni M, *et al.* 2012. Enhanced top soil carbon stocks under organic farming. *Proceedings of the National Academy of Sciences*, 109 (44): 18226–18231.

Ge Q, Wang H, Rutishauser T, *et al.* 2015. Phenological response to climate change in China: A meta-analysis. *Global Change Biology*, 21, 265–274.

Ge Q S, Zheng J Y, Hao Z X, *et al.* 2014. New progress in China's climate change research in the past 2000 years. *Acta Geographica Sinica*, 69, 1248–1258.

Godde M, Conrad R. 2000. Influence of soil properties on the turnover of nitric oxide and nitrous oxide by nitrification and denitrification at constant temperature and moisture. *Biology and Fertility of Soils*, 32 (2): 120–128.

He J, Li H W, Rasaily R G, *et al.* 2011. Soil properties and crop yields after 11 years of no-till farming in wheat-maize cropping system in North China Plain. *Soil & Tillage Research*, 113 (1): 48–54.

Huang S, Zeng Y, Wu J, *et al.* 2013. Effect of crop residue retention on rice yield in China: A meta-analysis. *Field Crops Research*, 154: 188–194.

Huang T, Yue X J, Ge X Z, *et al.* 2010. Evaluation of soil fertility quality in loess gully region based on principal component analysis: A case study of cultivated soil in Changwu Count. *Agricultural Research in Arid Areas*, 28(03):141-147+187.

IPCC, 2006. *IPCC Guidelines for National Greenhouse Gas Inventories*. Kanagawa, Japan: Institute for Global Environmental Strategies (IGES) for the IPCC.

IPCC, 2007. Climate change 2007: The physical science basis. In S Solomon, *et al.* (eds.) Contribution of Working Group I to the Fourth Assessment Report of the Intergovernmental Panel on Climate Change. Cambridge, UK and New York, NY, USA.

IPCC, 2012. Managing the risks of extreme events and disasters to advance climate change adaptation. A special report of working groups I and II of the intergovernmental panel on climate change.

IPCC, 2013. Climate change 2013: The physical science basis. In T F Stocker, D Qin, G K Plattner (eds.) Contribution of Working Group I to the Fifth Assessment Report of the Intergovernmental Panel on Climate Change. Cambridge, UK and New York, NY, USA.

IPCC, 2013. Climate Change 2013: The Physical Science Basis. Contribution of Working Group I to the Fifth Assessment Report of the Intergovernmental Panel on Climate Change. Cambridge University Press, Cambridge and New York, 1535.

IPCC, 2014. Climate change 2014: Synthesis report. Intergovernmental Panel on Climate Change, Geneva, Switzerland.

Lal R, Kimble J M. 1997. Conservation tillage for carbon sequestration. *Nutr Cycl Agroecosys*, 49:243–253.

Lal R, Follett R F, Stewart B, *et al.* 2017. Soil carbon sequestration to mitigate climate change and advance food security. *Soil Science*, 172 (12): 943–956.

Lal R, Kimble J, Levine E, *et al.* 1995. *Soils and Global Change*. CRC Press Inc. Environmental Science.

Lal R. 2004a. Agricultural activities and the global carbon cycle. *Nutrient Cycling in Agroecosystems*, 70 (2): 103–116.

Lal R. 2004b. Soil carbon sequestration impacts on global climate change and food security. *Science*, 304 (5677): 1623–1627.

Lal R. 2004c. Carbon emission from farm operations. *Environment International*, 30 (7): 981–990.

Lal R. 2005. World crop residues production and implications of its use as a biofuel. *Environment International*, 31 (4): 575–584.

Lal R. 2007. Anthropogenic Influences on World Soils and Implications to Global Food Security. Advances in Agronomy, 93: 69-93. 10.1016/S0065-2113(06) 93002-8.

Lal R. 2007. Soil Science and the Carbon Civilization. *Soil Science Society of America Journal*, 71: 1425-1437. https://doi.org/10.2136/sssaj2007.0001

Lal R. 2008. Crop residues as soil amendments and feedstock for bioethanol production. *Waste Management*, 28(4):747–758.

Lal R. 2009. Soil carbon sequestration to mitigate climate change. *Geoderma*, 123(1-2):1–22.

Lal R. 2010. Managing Soils and Ecosystems for Mitigating Anthropogenic Carbon Emissions and Advancing Global Food Security. *BioScience*, 60(9): 708–721.

Lal R. 2013. Intensive Agriculture and the Soil Carbon Pool. *Journal of Crop Improvement*, 27(6):735–751.

Lal R. 2015a. A system approach to conservation agriculture. *Journal of Soil and Water Conservation*, 70 (4): 82A-88A.

Lal R. 2015b. Sequestering carbon and increasing productivity by conservation agriculture. *Journal of Soil and Water Conservation*, 70 (3): 55A-62A.

Lal R. 2016a. Beyond COP 21: Potential and challenges of the "4 per Thousand" initiative. *Journal of Soil and Water Conservation*, 71 (1): 20A-25A.

Lal R. 2016b. Global food security and nexus thinking. *Journal of Soil and Water Conservation*, 71: 85A-90A.

Li C, Zhang Z, Guo L, *et al.* 2013. Emissions of CH_4 and CO_2 from double rice cropping systems under varying tillage and seeding methods. *Atmospheric Environment*, 80: 438–444.

Li H L, Wang C, Sun H T, et al. 2017. Research on Autonomous Obstacle Avoidance of Agricultural UAV Based on Deep Learning. *Agricultural Mechanization Research*, 39(08): 256–262.

Li T, Angeles O, Radanielson A, *et al.* 2015a. Drought stress impacts of climate change on rainfed rice in South Asia. *Climatic Change*, 133: 709–720.

Li T, Hasegawa T, Yin X, *et al.* 2015b. Uncertainties in predicting rice yield by current crop models under a wide range of climatic conditions. *Global Change Biology*, 21: 1328–1341.

Li X Y, Qin D H, Li J Y. 2007. *National Assessment Report on Climate Change*. Beijing: Science, 2007.

Li Y, Wu W, Ge Q, *et al.* 2016b. Simulating climate change impacts and adaptive measures for rice cultivation in Hunan Province, China. *Journal of Applied Meteorology and Climatology*, 55: 1359–1376.

Li Y M, Wang G L, Liu Z Y, et al. 2019. Effect of straw returning on the distribution characteristics of organic carbon and nitrate nitrogen in meadow soi. *Anhui Agricultural Sciences*, 47(24):63–66.

Li Z P, Liu M, Wu X C, *et al.* 2010. Effects of long-term chemical fertilization and organic amendments on dynamics of soil organic C and total N in paddy soilderived from barren land in subtropical China. *Soil & Tillage Research*, 106 (2): 268–274.

Li Z, Liu Z, Anderson W, *et al.* 2015c. Chinese rice production area adaptations to climate changes, 1949–2010. *Environmental Science and Technology*, 49: 2032–2037.

Liu B, Asseng S, Müller C, *et al.* 2016a. Similar estimates of temperature impacts on global wheat yield by three independent methods. *Nature Climate Change*, 6: 1130–1136.

Liu B, Liu L, Tian L *et al.* 2014c. Post-heading heat stress and yield impact in winter wheat of China. *Global Change Biology*, 20: 372–381.

Liu C, Lu M, Cui J, *et al.* 2014a. Effects of straw carbon input on carbon dynamics in agricultural soils: A meta-analysis. *Global Change Biology*, 20 (5): 1366–1381.

Liu L, Zhu Y, Tang L *et al.* 2013. Impacts of climate changes, soil nutrients, variety types and management practices on rice yield in East China: A case study in the Taihu region. *Field Crops Research*, 149: 40–48.

Liu S L, Huang D Y, Chen A L, *et al.* 2014b. Differential responses of crop yields and soil organic carbon stock to fertilization and rice straw incorporation in three cropping systems in the subtropics. *Agriculture Ecosystems & Environment*, 184: 51–58.

Liu S, Pu C, Ren Y, *et al.* 2016b. Yield variation of double-rice in response to climate change in Southern China. *European Journal of Agronomy*, 81: 161–168.

Liu Z J, Yang X G, Hubbard K G, *et al.* 2012. Maize potential yields and yield gaps in the changing climate of northeast China. *Global Change Biology*, 18 (11): 3441–3454.

Liu Z, Yang P, Tang H, *et al.* 2015. Shifts in the extent and location of rice cropping areas match the climate change pattern in China during 1980–2010. *Regional Environmental Change*, 15, 919–929.

Lobell D B, Banziger M, Magorokosho C, *et al.* 2011a. Nonlinear heat effects on African maize as evidenced by historical yield trials. *Nature Climate Change*, 1, 42–45.

Lobell D B, Burke M B, Tebaldi C, *et al.* 2008. Prioritizing climate change adaptation needs for food security in 2030. *Science*, 319: 607–610.

Lobell D B, Hammer G L, Chenu K, *et al.* 2015a. The shifting influence of drought and heat stress for crops in northeast Australia. *Global Change Biology*, 21: 4115–4127.

Lobell D B, Roberts M J, Schlenker W, *et al.* 2015b. Greater sensitivity to drought accompanies maize yield increase in the U. S. Midwest. *Science*, 2014, 344: 516–519.

Lobell D B, Schlenker W, Costa-Roberts J. 2011b. Climate trends and global crop production since 1980. *Science*, 333 (6042): 616–620.

Lu F, Wang X, Han B, *et al.* 2009. Soil carbon sequestrations by nitrogen fertilizer application, straw return and no-tillage in China's cropland. *Global Change Biology*, 15 (2): 281–305.

Mangalassery S, Sjogersten S, Sparkes D L, *et al.* 2014. To what extent can zero tillage lead to a reduction in greenhouse gas emissions from temperate soils. *Scientific Report*, 4: 4586.

Menzel A. 2000. Trends in phenological phases in Europe between 1951 and 1996. *International Journal of Biometeorology*, 44: 76–81.

Moore F C, Diaz D B. 2015. Temperature impacts on economic growth warrant stringent mitigation policy. *Nature Climate Change*, 5: 127–131.

Olesen J E, Trnka M, Kersebaum K C, *et al.* 2011. Impacts and adaptation of European crop production systems to climate change. *European Journal of Agronomy*, 34: 96–112.

Palm C, Blanco-Canqui H, DeClerck F, *et al.* 2014. Conservation agriculture and ecosystem services: An overview. *Agriculture, Ecosystems & Environment*, 187: 87–105.

Piao S, Ciais P, Huang Y, *et al.* 2010. The impacts of climate change on water resources and agriculture in China. *Nature*, 467: 43–51.

Pittelkow C M, Fischer A J, Moechnig M J, *et al.* 2012. Agronomic productivity and nitrogen requirements of alternative tillage and crop establishment systems for improved weed control in direct-seeded rice. *Field Crops Research*, 130: 128–137.

Pittelkow C M, Liang X, Linquist B A, *et al.* 2015a. Productivity limits and potentials of the principles of conservation agriculture. *Nature*, 517 (7534): 365–368.

Pittelkow C M, Linquist B A, Lundy M E, *et al.* 2015b. When does no-till yield more? A global Meta-analysis. *Field Crops Research*, 183: 156–168.

Qin D H. 2004. The science of climate change entering the 21st century: Facts, effects and countermeasures of climate change. *Science and Technology Review*, 22: 4–7.

Qin D H. 2014. Climate change science and human sustainable development. *Advances in Geographical Sciences*, 33: 874–883.

Ray D K, Gerber J S, MacDonald G K, *et al.* 2015. Climate variation explains a third of global crop yield variability. *Nature Communications*, 6: 5989.

Reeves D W. 1997. The role of soil organic matter in maintaining soil quality in continuous cropping systems. *Soil & Tillage Research*, 43 (1–2): 131–167.

Rezaei E E, Siebert S, Ewert F. 2017. Climate and management interaction cause diverse crop phenology trends. *Agricultural and Forest Meteorology*, 233: 55–70.

Rosenzweig C, Elliott J, Deryng D, *et al.* 2014. Assessing agricultural risks of climate change in the 21st century in a global gridded crop model intercomparison. *Proceedings of the National Academy of Sciences of the United States of America*, 111: 3268–3273.

Ruan L L, Robertson G P. 2013. Initial nitrous oxide, carbon dioxide, and methane costs of converting conservation reserve program grassland to row crops under no-till vs. conventional tillage. *Global Change Biology*, 19 (8): 2478–2489.

Sá, J C D, Lal, R. 2009. Stratification ratio of soil organic matter pools as an indicator of carbon sequestration in a tillage chronosequence on a Brazilian Oxisol. *Soil & Tillage Research*, 103 (1): 46–56.

Sacks W J, Kucharik C J. 2011. Crop management and phenology trends in the U. S. Corn Belt: Impacts on yields, evapotranspiration and energy balance. *Agricultural and Forest Meteorology*, 151, 882–894.

Siebert S, Ewert F. 2012. Spatio-temporal patterns of phenological development in Germany in relation to temperature and day length. *Agricultural and Forest Meteorology*, 152: 44–57.

Singh K, Mcclean C J, Buker P *et al.* 2017a. Mapping regional risks from climate change for rainfed rice cultivation in India. *Agricultural Systems*, 156: 76–84.

Singh V, Nguyen C T, Mclean G, *et al.* 2017b. Quantifying high temperature risks and their potential effects on sorghum production in Australia. *Field Crops Research*, 211: 77–88.

Six J, Feller C, Denef K, *et al.* 2002. Soil organic matter, biota and aggregation in temperate and tropical soils — Effects of no-tillage. *Agronomie*, 22 (7–8): 755–775.

Six J, Ogle S M, Jay breidt F, *et al.* 2004. The potential to mitigate global warming with no-tillage management is only realized when practised in the long term. *Global Change Biology*, 10 (2): 155–160.

Skinner C, Gattinger A, Muller A, *et al.* 2014. Greenhouse gas fluxes from agricultural soils under organic and non-organic management — A global Meta-analysis. *Science of the Total Environment*, 468–469: 553–563.

Smith K A, Conen F. 2004. Impacts of land management on fluxes of trace greenhouse gases. *Soil Use and Management*, 20: 255–263.

Smith P, Martino D, Cai Z, *et al.* 2008. Greenhouse gas mitigation in agriculture. *Philosophical Transactions of the Royal Society B-Biological Sciences*, 363 (1492): 789–813.

Tao F, Hayashi Y, Zhang Z, *et al.* 2008. Global warming, rice production, and water use in China: Developing a probabilistic assessment. *Agricultural and Forest Meteorology*, 148: 94–110.

Tao F, Zhang S, Zhang Z, *et al.* 2014. Maize growing duration was prolonged across China in the past three decades under the combined effects of temperature, agronomic management, and cultivar shift. *Global Change Biology*, 20: 3686–3699.

Tao F, Zhang Z, Shi W, *et al.* 2013. Single rice growth period was prolonged by cultivars shifts, but yield was damaged by climate change during 1981–2009 in China, and late rice was just opposite. *Global Change Biology*, 19: 3200–3209.

Thomas D T, Lawes R A, Descheemaeker K, *et al.* 2014. Selection of crop cultivars suited to the location combined with astute management can reduce crop yield penalties in pasture cropping systems. *Crop & Pasture Science*, 65: 1022–1032.

Tilman D, Balzer C, Hill J, *et al.* 2011. Global food demand and the sustainable intensification of agriculture. *Proceedings of the National Academy of Sciences of the United States of America*, 108 (50): 20260–20264.

United Nations, Department of Economic and Social Affairs, Population Division. 2014.World Urbanization Prospects: The 2014 Revision, Methodology Working Paper No. ESA/P/WP.238

United Nations, Department of Economic and Social Affairs, Population Division. 2019. Report of the Secretary-General on the Work of the Organization, Methodology Working Paper.

Ussiri D A N, Lal R, Jarecki M K. 2009. Nitrous oxide and methane emissions from long-term tillage under a continuous corn cropping system in Ohio. *Soil & Tillage Research*, 104 (2): 247–255.

Ussiri D A N, Lal R. 2009. Long-term tillage effects on soil carbon storage and carbon dioxide emissions in continuous corn cropping system from an alfisol in Ohio. *Soil & Tillage Research*, 104 (1): 39–47.

Van Bussel L G J, Stehfest E, Siebert S, *et al.* 2015. Simulation of the phenological development of wheat and maize at the global scale. *Global Ecology and Biogeography*, 24: 1018–1029.

Van Kessel C, Venterea R, Six J, *et al.* 2013. Climate, duration, and N placement determine N_2O emissions in reduced tillage systems: A meta-analysis. *Global Change Biology*, 19 (1): 33–44.

Wang B, Liu L, O'leary G J, *et al.* 2017a. Australian wheat production expected to decrease by the late 21st century. *Global Change Biology*, 24(6): 2403–2415.

Wang J, Yang Y, Huang J, *et al.* 2015a. Information provision, policy support, and farmers' adaptive responses against drought: An empirical study in the North China Plain. *Ecological Modelling*, 318: 275–282.

Wang Y D, Hu N, Xu M G, *et al.* 2015b. 23-year manure and fertilizer application increases soil organic carbon sequestration of a rice-barley cropping system. *Biology and Fertility of Soils*, 51 (5): 583–591.

Wang Y Y, Li X X, Dong W X, et al. 2018. Review on greenhouse gas emission and reduction in wheat-maize double cropping system in the North China Plain. *Chinese Journal of Eco-Agriculture*, 26(02): 167–174.

Wang Z, Shi P, Zhang Z, *et al.* 2017b. Separating out the influence of climatic trend, fluctuations, and extreme events on crop yield: A case study in Hunan Province, China. *Climate Dynamics*, 1–19.

Wei Y H, Zhao X, Zhai Y L, *et al.* 2013. Effects of farming methods on soil carbon sequestration in North China. *Transactions of the Chinese Society of Agricultural Engineering*, 29 (17): 87–95.

West T O, Marland G. 2002. A synthesis of carbon sequestration, carbon emissions, and net carbon flux in agriculture: Comparing tillage practices in the United States. *Agriculture Ecosystems & Environment*, 91 (1–3): 217–232.

West T O, Post W M. 2002. Soil organic carbon sequestration rates by tillage and crop rotation. *Soil Science Society of America Journal*, 66 (6): 1930–1946.

West T O, Six J. 2007. Considering the influence of sequestration duration and carbon saturation on estimates of soil carbon capacity. *Climatic Change*, 80 (1–2): 25–41.

Xie R Z, Li S K, Jin Y Z, *et al.* 2008. Yield effect analysis of experimental research on conservation tillage in China. *Chinese Agricultural Sciences*, 41 (2): 397–404.

Xiong W, Conway D, Lin E, *et al.* 2009a. Potential impacts of climate change and climate variability on China's rice yield and production. *Climate Research*, 40: 23–35.

Xiong W, Conway D, Lin E, *et al.* 2009b. Future cereal production in China: The interaction of climate change, water availability and socio-economic scenarios. *Global Environmental Change-Human and Policy Dimensions*, 19: 34–44.

Xiong W, Holman I, Lin E D, *et al.* 2010. Climate change, water availability and future cereal production in China. *Agriculture Ecosystems & Environment*, 135: 58–69.

Xiong W, Van Der Velde M, Holman I P, *et al.* 2014. Can climate-smart agriculture reverse the recent slowing of rice yield growth in China? *Agriculture Ecosystems & Environment*, 196: 125–136.

Xue J F, Liu S L, Chen Z D, *et al.* 2014. Assessment of carbon sustainability under different tillage systems in a double rice cropping system in southern China. *The International Journal of Life Cycle Assessment*, 19 (9): 1581–1592.

Xue J F, Zhao X, Chen F, *et al.* 2013. Research progress on the effects of conservation tillage on farmland carbon and nitrogen effects. *Acta Ecologica Sinica*, 33 (19): 6006–6013.

Yang X G, Liu Z J, Chen F. 2010. The possible impact of global warming on China's cropping system I. An analysis of the possible impact of climate warming on the northern boundary of China's cropping system and grain production. *Chinese Agricultural Sciences*, 43 (2): 329–336

Yang X, Chen F, Lin X, *et al.* 2015. Potential benefits of climate change for crop productivity in China. *Agricultural and Forest Meteorology*, 208, 76–84.

Yao F M. 2005. *Evaluation of the Impact of Climate Change on My Country's Grain Production: Taking Rice as an Example*. Beijing: University of Chinese Academy of Sciences.

Yao Z, Zheng X, Wang R, *et al.* 2013. Nitrous oxide and methane fluxes from a rice-wheat crop rotation under wheat residue incorporation and no-tillage practices. *Atmospheric Environment*, 79: 641–649.

Yu F W, Yang G. 2018. Current situation, dilemma and countermeasures of resource utilization of crop straw. *Social Scientist*, 2:33–39.

Zhang F S, Chen X P, Vitousek P. 2013d. An experiment for the world. *Nature*, 497 (7447): 33–35.

Zhang H L, Bai X L, Xue J F, *et al.* 2013e. Emissions of CH_4 and N_2O under different tillage systems from double-cropped paddy fields in southern China. *Plos One*, 8 (6): e65277.

Zhang H L, Lal R, Zhao X, *et al.* 2014c. Opportunities and challenges of soil carbon sequestration by conservation agriculture in China. *Advances in Agronomy*, 124: 1–36.

Zhang M Y, Wang F J, Chen F, *et al.* 2013c. Comparison of three tillage systems in the wheat-maize system on carbon sequestration in the North China Plain. *Journal of Cleaner Production*, 54: 101–107.

Zhang S, Tao F L, Zhang Z. 2014a. Rice reproductive growth duration increased despite of negative impacts of climate warming across China during 1981–2009. *European Journal of Agronomy*, 54: 70–83.

Zhang S, Tao F, Zhang Z. 2016a. Changes in extreme temperatures and their impacts on rice yields in southern China from 1981 to 2009. *Field Crops Research*, 189: 43–50.

Zhang S, Tao F, Zhang Z. 2017a. Uncertainty from model structure is larger than that from model parameters in simulating rice phenology in China. *European Journal of Agronomy*, 87: 30–39.

Zhang T, Huang Y, Yang X. 2013a. Climate warming over the past three decades has shortened rice growth duration in China and cultivar shifts have further accelerated the process for late rice. *Global Change Biology*, 19: 563–570.

Zhang T, Huang Y. 2012. Impacts of climate change and inter-annual variability on cereal crops in China from 1980 to 2008. *Journal of the Science of Food and Agriculture*, 92: 1643–1652.

Zhang T, Li T, Yang X, *et al.* 2016b. Model biases in rice phenology under warmer climates. *Scientific Reports*, 6. 27355 (https://doi.org/10.1038/srep27355).

Zhang T, Simelton E, Huang Y, *et al.* 2013b. A Bayesian assessment of the current irrigation water supplies capacity under projected droughts for the 2030s in China. *Agricultural and Forest Meteorology*, 178–179: 56–65.

Zhang T, Yang X, Wang H, *et al.* 2014b. Climatic and technological ceilings for Chinese rice stagnation based on yield gaps and yield trend pattern analysis. *Global Change Biology*, 20: 1289–1298.

Zhang T, Yang X. 2016. Mapping Chinese rice suitability to climate change. *Journal of Agricultural Science*, 8: 33.

Zhang X Q, Pu C, Zhao X, *et al.* 2016d. Tillage effects on carbon footprint and ecosystem services of climate regulation in a winter wheat-summer maize cropping system of the North China Plain. *Ecological Indicators*, 67: 821–829.

Zhang Y, Zhao Y, Wang C, *et al.* 2016c. Using statistical model to simulate the impact of climate change on maize yield with climate and crop uncertainties. *Theoretical and Applied Climatology*, 130: 1065–1071.

Zhang Z, Chen Y, Wang C, *et al.* 2017b. Future extreme temperature and its impact on rice yield in China. *International Journal of Climatology*, 37: 4814–4827.

Zhang W, Hou L B, Zhang B, et al. 2005. Comparative study on ridge cropping and conservation tillage methods in western Liaoning. *Chinese Agricultural Science Bulletin*, 07:175-178.

Zhao C, Liu B, Piao S, *et al.* 2017a. Temperature increase reduces global yields of major crops in four independent estimates. *Proceedings of the National Academy of Sciences of the United States of America*, 114: 9326–9331.

Zhao C, Piao S, Wang X, *et al.* 2016a. Plausible rice yield losses under future climate warming. *Nature Plants*, 3: 16202.

Zhao H, Fu Y H, Wang X, *et al.* 2016b. Timing of rice maturity in China is affected more by transplanting date than by climate change. *Agricultural and Forest Meteorology*, 216, 215–220.

Zhao X, Liu S L, Pu C, *et al.* 2017b. Crop yields under no-till farming in China: A meta-analysis. *European Journal of Agronomy*, 84: 67–75.

Zhao X, Xue J F, Zhang X Q, *et al.* 2015a. Stratification and storage of soil organic carbon and nitrogen as affected by tillage practices in the North China Plain. *Plos One*, 10 (6): e0128873.

Zhao X, Zhang R, Xue J F, *et al.* 2015b. Management-induced changes to soil organic carbon in China: A meta-analysis. *Advances in Agronomy*, 134: 1–50.

Zou J, Huang Y A O, Qin Y, *et al.* 2009. Changes in fertilizer-induced direct N_2O emissions from paddy fields during rice-growing season in China between 1950s and 1990s. *Global Change Biology*, 15 (1): 229–242.

Chapter 7

Practice and Exploration of Wheat–Rice Production with Climate-Smart Agriculture in China

Abstract

Climate-smart food crop production project was implemented in the Huaiyuan County, Anhui Province, from 2015 to 2019. This project aimed at technical practices, technical training, service consultations, monitoring, evaluation, and other works of wheat–rice production system. During the implementation stage, the project team made phased achievements in terms of reduction in greenhouse gas emissions, high yield and high efficiency of crops by exploring the impact of different new materials of carbon sequestration and emission reduction, new modes and supporting cultivation techniques of conservation tillage on farmland in wheat–rice production system. At the same time, the project team conducted systematic training on wheat–rice production for local farmers, agricultural machinery personnel, village cadres and other personnel, not only carrying out a large number of lectures on "climate-smart agriculture" but also providing technical guidance and demonstration of pest control on local production. A new combination technology of planting and breeding and the teaching of new knowledge and new content in supporting sales of agricultural products encourages the rural areas and farmers in the project area embark on real rural revitalization. Production monitoring and evaluation of climate-smart wheat–rice system promotes the comparative effects of carbon sequestration and emission reduction, environmental effects, pest management, and social impact of the production system under project conditions and baseline conditions. The experimental results proved that the effects of increasing production and reducing emissions were based upon the China's climate-smart agricultural wheat–rice system, and at the same time, it could improve water quality, better prevent and control farmland pests and diseases in the demonstration area, reduce the cost of medicine, and increase the effective utilization of pesticides. The project results have achieved the expected results. Farmers were able to gain further understanding and acceptance of the concepts, technologies, and policies of climate-smart agriculture and had a strong willingness to apply them into production, which strongly promotes the development of climate-smart agriculture in the winter wheat–rice system application.

7.1 Practice of Climate-Smart Wheat–Rice Production Technology

In China's climate-smart agricultural wheat–rice production system, the focus is on the release of different new materials for carbon sequestration and emission reduction, new patterns of carbon sequestration and emission reduction, and technologies for conservation tillage, focusing on greenhouse gas emissions reduction, high-yield, and efficient coordinated development of crops. The experimental results of the effect of different new materials for carbon sequestration and emission reduction on the greenhouse gas emissions and yield of rice and wheat production showed that in the annual greenhouse gas emissions, CH_4 emissions mainly depended on the rice season, and the total N_2O emissions during the wheat season accounted for a larger share of the total annual N_2O emissions. Fertilization and straw burning increase the annual global warming potential of the wheat–rice system. Under the condition of conventional fertilization and full return of straw, adding nitrification inhibitors and biochar or ammonium sulfate could effectively reduce the annual warming potential of the wheat–rice system. Both of the rice season and the wheat season, increased application of nitrogen fertilizer could increase crop yield, and increased application of nitrification inhibitors could also promote the increase in yield. In the wheat season, the combination of nitrification inhibitors and ammonium sulfate could increase yield, and the conventional fertilization for rice seasons with nitrification inhibitors and biochar had a better effect on the yield increase.

The experimental results of new patterns of carbon sequestration and emission reduction in different farmlands showed that compared with the rice–wheat rotation pattern, which has the highest greenhouse gas emissions, the three patterns of rice–rape, rice–green manure, and rice–winter fallow all have the effect of reducing greenhouse gas emissions, and the winter fallow system has the best emission reduction effect. A comprehensive comparison of the rice yield and winter crop biomass of different new carbon sequestration and emission reduction patterns showed that the winter crop biomass of the rice–tallow vetch pattern was the largest and had the best effect on increasing rice production. It could be seen that planting rapeseed, green manure, or winter fallow in winter crops were all alternatives to the new pattern of carbon sequestration and emission reduction instead of the rice–wheat production system. Considering the effects of carbon sequestration and emission reduction, winter crop biomass, and the rice yield, the rice–green manure pattern was the most suitable planting pattern.

The experimental results on the impact of different supporting cultivation techniques of conservation tillage on the rice–wheat production system showed that the plowing and rotary tillage measures in the rice season had the N_2O reducing effect in both the rice and wheat seasons, and it could effectively reduce the system's N_2O emissions. In the rice–wheat production system, different soil cultivation methods had greater impacts on the greenhouse gas emissions of the wheat season. Compared with no-tillage, rotary tillage of wheat season reduced N_2O emissions and increased CH_4 emissions, but CH_4 contributed less to greenhouse gas emissions per unit area, so rotary tillage of wheat season was beneficial to reducing Global Warming Potential (GWP) and Greenhouse Gas Intensity (GHGI). Regardless of the rice season or the wheat season, returning the full amount of straw to the field could reduce N_2O and CH_4 emissions during the growth period of the crop. Under the condition that all wheat straws were returned to the field, the greenhouse gas emission reduction effect during the wheat season combined with rotary tillage farming measures was better. Under the same condition, appropriately increasing the intensity of tillage measures (such as rotary tillage) was conducive to stable yield, but the effect of rotary tillage and tillage measures in the rice season on rice yield was not obvious.

7.1.1 *Screening and Demonstration of New Materials for Carbon Sequestration and Emission Reduction in Farmland*

The Huaiyuan project focuses on the rice–wheat rotation system to explore the impact of different new materials for carbon sequestration and emission reduction on the greenhouse gas emissions and yield of rice and wheat production (Figure 7.1). Based on the 2015–2019 experiment, it focused on measuring the emission rate and cumulative emissions of greenhouse gases such as CH_4 and N_2O from different new carbon sequestration and emission reduction materials, as well as the warming potential and greenhouse gas emission intensity of the wheat and rice production system. Researchers systematically analyzed the effects of different experimental treatments on greenhouse gas emissions from the production system and studied the effects of different experimental treatments on wheat and rice output and yield components factors. The study results showed that in the annual wheat–rice production system, no fertilization

Figure 7.1.   Screening and demonstration of new materials for carbon sequestration and emission reduction in wheat–rice project area.

treatment in the rice season increased CH_4 emissions from rice fields and reduced the total emissions of N_2O. Increased application of nitrogen fertilizers and new material ammonium sulfate increased the rate of N_2O emissions, thereby increasing the rate of N_2O emissions, but it had a reduction effect on CH_4, and the reduction effect on CH_4 was better when the full amount of straw was returned to the field. Among these new materials, ammonium sulfate had the most significant reduction effect on CH_4 emissions. Nitrogen fertilization and straw burning in the wheat season increased greenhouse gas emissions, while adding nitrification inhibitors and returning straw to the field could reduce farmland greenhouse gas emissions, and the emission reduction effect of adding nitrification inhibitors under the condition of full straw return was the most significant.

(1) Influence of different emission reduction materials on greenhouse gas emissions of wheat–rice production system
(i) CH_4 and N_2O emission dynamics from rice production: During the rice season in a 2015 experiment, five treatments were set up: no fertilization treatment (CK), new material ammonium sulfate treatment (NM), no fertilization treatment under the condition of full straw returning (CK+S), conventional fertilization treatment under the condition of full straw returning (FP+S), and increased ammonium sulfate treatment under the condition of full straw returning (NM+S). The emission rate of CH_4 in paddy field decreased with the increasing application of ammonium sulfate, and the emission rate of CH_4 in growth period was more stable than that without fertilization. Taking the peak seedling period (September 2) as the dividing line, before the peak, the CH_4 emission rate of the full amount of straw returned to the field without fertilization was higher. After the peak, the CH_4 emission rate of the control group without fertilization began to increase rapidly, and the CH_4 emission rate reached its maximum within a certain time. The new material ammonium sulfate treatment effectively reduced the CH_4 emission rate, and the effect was more significant under the condition of full straw return. Although conventional fertilization could effectively reduce the CH_4 emission rate, its emission reduction effect was not significant compared with the new material ammonium sulfate.

In 2015, the effects of each treatment in the rice season on the N_2O emission rate showed a single-peak change trend. There were little difference between treatments from transplanting to August 5. From August 5 to 26, the emission peak of each treatment appeared. This was because in the middle and late August, rice tillering ended and rice production and management entered the shelf period. At this time, almost no water layer existed in the paddy field, the water content in the field was relatively low, and the N_2O emission rate increased and appeared at a peak. At this point, the emission rate of the total amount of the new material straw returning to the field (NM+S) was the highest, and the emission rate of the control group without fertilization (CK) was the lowest. Compared with no fertilization treatment (CK), before August 20, the new material ammonium sulfate treatment (NM), no fertilization treatment under the condition of full straw returning (CK+S), conventional fertilization treatment under the condition of full straw returning (FP+S), and increased ammonium sulfate treatment under the condition of full straw returning (NM+S) increased the N_2O emission rate by approximately 61.7%, 43.7%, 69.5%, and 114.4%, respectively. After August 20, the N_2O emission rate of the treatment with full amount of straw returned to the field was higher than that of the treatment without returning to the field. The experiments results showed that the interaction between the added nitrogen fertilizer and the full amount of straw returned to the field significantly increased the N_2O emission rate, and increased acidity fertilizers caused the largest increase in N_2O emissions.

The experimental results in 2015 showed that the control group without fertilization treatment (CK) significantly increased the total CH_4 emissions from the rice fields while reducing the total N_2O emissions from the rice fields; increasing nitrogen fertilizer application increased the total N_2O emissions from the rice fields and reduced the total CH_4 emissions from the rice fields. Under the condition of returning straw to the field, adding nitrogen fertilizer had a more significant effect on the increase of total N_2O emissions and the reduction of total CH_4 emissions.

(ii) CH_4 and N_2O emission dynamics from wheat production: In 2016 and 2017, six treatments were set up in the wheat season experiments: no fertilization treatment (CK), conventional fertilization (CF), conventional fertilization + full straw return to the field (CFS), conventional fertilization + straw burning to the field (CF+BS), conventional fertilization of straw without returning to the field + nitrification inhibitor (CF+DCD) and conventional fertilization of full straw returning to the field + nitrification inhibitor (CFS+DCD). The study results showed that CH_4 emission in the wheat season was much smaller than it in the rice season, and the emission rate of each treatment tended to zero on April 13, but the emission trend of each treatment was not obvious. Compared with other treatments, the CH_4

emission rate under no fertilization treatment (CK) was the smallest, and the emission rate of conventional fertilization + full straw return to the field (CFS) reached its peak on January 14 and reached its emission second peak on March 1, but overall, there was little difference between the treatments, and there was no significant trend of change.

In 2016, there were two peaks of N_2O emissions during the wheat growth period: The first was in mid-February when the temperature rose gradually and the base fertilizer applied in the previous period was gradually used by the plants. The second was in mid-March. This was because the additional fertilizer was applied to wheat at this time and the input of nitrogen fertilizer accelerated the production of N_2O, but because no fertilization treatment (CK) did not apply fertilizer, the N_2O emission was low. Adding nitrification inhibitors could effectively reduce the N_2O emission rate. Different straw returning methods had different effects on the N_2O emission rate. Under conventional fertilization conditions, the N_2O emission rate of the straw burning to the field (CF+BS) was significantly greater than that under the all-straw return to the field (CFS). So, from the perspective of N_2O emission reduction, we should vigorously promote the full return of straw to the field to reduce straw burning. The N_2O emission trend in the 2017 wheat season was consistent with that in the 2016 wheat season, and there were also two peaks. The first time was in mid-February, and the second time occurred in mid-April, because the top-dressing period was delayed in the 2016 wheat season. Compared with conventional fertilization, regardless of whether the straw was returned to the field or not, the N_2O emission rate in the early stage of wheat growth was significantly reduced by adding nitrification inhibitor. However, in the later growth period (early May), as the effect of the nitrification inhibitor decreases, conventional fertilization of straw without returning to the field+nitrification inhibitor (CF+DCD) and conventional fertilization of full straw returning to the field+nitrification inhibitor (CFS+DCD) gradually showed the emission peak, which was later than the conventional fertilization treatment and the control. Therefore, increasing the application rate of nitrification inhibitors has a greater inhibitory effect on N_2O emissions in the early growth period of wheat, while the emission reduction effect in the later growth period is weak.

The experimental results during the 2016 wheat season showed that under the condition of no straw returning, the N_2O emission without fertilization was 350 kg CO_2-eq hm^{-2}, and that under the conventional fertilization treatment increased by 263.4%, but it decreased by about 88%

after adding the nitrification inhibitor; Compared with no returning, the N_2O emission of returning all straw under conventional fertilization was reduced by about 17.6%, and the N_2O emission dropped by 17.6% after the application of nitrification inhibitors. The results of the study showed that adding nitrification inhibitors had a significant effect on reducing N_2O emissions; returning straw to the field could reduce N_2O emissions and increase CH_4 emissions. Compared with conventional fertilization, CH_4 emissions increased slightly after adding the nitrification inhibitor, which might be related to the increase in soluble organic carbon content in the field by adding the nitrification inhibitor.

The greenhouse gas emissions in the 2017 wheat season experiment were the same as that in 2016, the N_2O and CH_4 emissions of the treatment without fertilization under the condition of straw returning to the field were the smallest, and conventional fertilization (CF) and conventional fertilization under the conditions of returning straw to the field (CFS) both increased the emissions of CH_4 and N_2O, and adding nitrification inhibitors could significantly reduce N_2O emissions, but might also increase CH_4 emissions. The study results showed that conventional fertilization and straw burning to the field significantly increased N_2O and CH_4 emissions from wheat. The 2-year experiment results showed that returning all straws to the field could help reduce greenhouse gas emissions from farmland compared to straw burning to the field.

(iii) Annual greenhouse gas emissions of wheat–rice system: In 2016 and 2017, eight groups of rice–wheat annual rotation experiments were set up: T1 (no fertilization treatment), T2 (conventional fertilization treatment), T3 (add nitrification inhibitor and biochar under conventional fertilization treatment), T4 (add nitrification inhibitor and ammonium sulfate under conventional fertilization treatment), T5 (full straw returned to the field under conventional fertilization conditions), T6 (full straw returned to the field and increased application of nitrification inhibitor and biochar under conventional fertilization conditions), T7 (full straw returned to the field and increased application of nitrification inhibitor and ammonium sulfate under conventional fertilization conditions), and T8 (straw burning to the field under conventional fertilization conditions).

The annual CH_4 emissions of the wheat–rice crop rotation system mainly depended on the rice season, and the proportion of CH_4 emissions in the wheat season was very small. Under the condition of returning all straws to the field, both conventional fertilization and increased application of nitrification inhibitors reduced CH_4 emissions. With the addition of nitrification inhibitors, the emissions of CH_4 in 2017 were significantly reduced, and the effect of reducing emissions in 2016 was not obvious. Compared with conventional fertilization treatments, the increased application of biochar and ammonium sulfate in 2016 reduced the CH_4 emissions from paddy fields in the rice season. Among them, the reduction effect of biochar was more obvious, and the difference between treatments in 2017 was not significant. In addition, the study results also reflected large inter-annual differences. We speculated that this had a greater relationship with the cumulative effects of the processing settings and changes of the inter-annual meteorological factors.

The annual N_2O emissions of the wheat–rice rotation system are mainly determined by the wheat season, and the N_2O emissions of the wheat season account for a relatively large amount of the total annual N_2O emissions. Compared with the control group, increasing nitrogen fertilizer application increased the N_2O emissions in the wheat season but reduced the N_2O emissions in the rice season, and the annual N_2O emission was increased. Compared with conventional fertilization, adding nitrification inhibitors under the treatment of straws not returning to the field could significantly reduce N_2O emissions in the wheat season. However, compared with conventional fertilization (T5), under the treatment of returning straw to the field, only full straw returned to the field and increased application of nitrification inhibitor and biochar fertilization under conventional conditions (T6) reduced N_2O emissions in the wheat season by about 11.7%, but full straw returned to the field and increased application of nitrification inhibitor and ammonium sulfate under conventional fertilization conditions (T7) in the wheat season increased N_2O by about 35%. This might be because the increased application of ammonium sulfate in the rice season would change the physical and chemical properties of the soil in the wheat season, and increased substrates of soil nitrification and denitrification finally led to an increase in N_2O emissions in the wheat season. In each treatment group, under the condition of conventional fertilization, full straw returned to the field and increased application of nitrification inhibitor and ammonium sulfate under conventional fertilization conditions (T7) had the highest N_2O emissions. From an annual perspective, under the full amount of straw returned to the field, the increased application of nitrification inhibitor in wheat season plus the increased application of biochar in rice season were both beneficial to the reduction of N_2O emissions.

It is clear that based on the effects of wheat and rice on the global warming potential under the use of different emission reduction materials, compared with the treatment without fertilization (T1), increasing nitrogen fertilizer could increase the annual GWP. There was no significance within treatments in 2016, but it was remarkable in 2017. In the case of straws not returning to the field, adding nitrification inhibitors reduced the annual global warming potential, but the differences between treatments were not significant. When the full straw returned to the field, the annual global warming potential of the straw burning to the field under conventional fertilization conditions (T8) was the largest. Compared with conventional fertilization, the GWP of the T6 and T7 treatments in 2017 decreased by approximately 16.4% and 4.1%, respectively, but the difference between the treatments in the 2016 wheat season was not significant.

Comparing the effects of different emission-reducing treatments on greenhouse gas emissions per unit yield of wheat and rice, the results showed that under the treatment of straws not returning to the field, adding nitrification inhibitors could reduce the greenhouse gas emissions per unit yield of wheat and rice, and adding biochar had a significant difference in the 2017 wheat season. Compared with conventional fertilization, in the case of returning all straw to the field, adding nitrification inhibitors and biochar could reduce greenhouse gas emissions per unit yield of wheat and rice, but the effects of the treatments with adding ammonium sulfate were not significant.

(2) Influence of different emission reduction materials on greenhouse gas emissions of wheat–rice yield

(i) The impact of different emission reduction materials on single-season rice and wheat yields: The experimental results in 2015 showed that under the condition of not returning straw to the field, the yields of rice with and without fertilization (CK) were 12.1 and 11.7 t hm^{-2}, respectively. Adding ammonium sulfate under the condition of not returning to the field (NM), conventional fertilization under the condition of returning full straw to the field (FP+S) and the increased application of ammonium sulfide under the condition of returning full straw to the field (NM+S), rice yields were 13.6, 13.3 and 14.9 t hm^{-2}, respectively. Under the condition of returning the full straw to the field, the effect of increasing the yield of rice with added ammonium sulfate was the most significant. The experimental results in 2016 showed that no fertilization treatment (CK), conventional fertilization (CF), conventional fertilization + full return of straw to the field (CFS), conventional fertilization + straw burning to the field (CF+BS), conventional fertilization without returning straw to the field + nitrification inhibitor (CF+DCD), and conventional fertilization with full return of straw + nitrification inhibitor treatment (CFS+DCD), the wheat yields were 5.5, 5.4, 5.6, 5.8, 5.6 and 5.6 t hm^{-2}, respectively, and the wheat yield was the highest under the condition of conventional fertilization + straw burning to the field (CF+BS), but there was no significant difference among the treatments.

(ii) Effects of different new materials for emission reduction on the annual yield of rice–wheat: The experimental results showed that rice yield accounted for a relatively large percentage of annual crop yields in the rice–wheat rotation system. Compared with the control treatment (CK) without fertilization, increased application of nitrogen fertilizer (CF) could increase crop yield. In the wheat season, no matter the straw was returned to the field or not, the increased application of nitrification inhibitors increased the yield of wheat, and the yield of treatments with additional nitrification inhibitors and ammonium sulfate under conventional fertilization increased the most compared with the control group (CK). Compared with conventional fertilization in the rice season, the application of nitrification inhibitors could increase crop yields, and the application of nitrification inhibitors and biochar under conventional fertilization in the rice season had better yield-increasing effects. Regarding the annual crop yield, compared with conventional fertilization, the annual yield increased by approximately 12.6% by adding nitrification inhibitors (dicyandiamide) in the wheat season and adding biochar in the rice season when returning full straw to the field. The effect of straw burning to field on crop yield was not obvious.

7.1.2 *Screening and Demonstration of New Modes for Carbon Sequestration and Emission Reduction in Farmland*

Since November 2015, the Huaiyuan area has experimented with four types of crop rotation mode treatments, including rice–wheat (RW), rice–oilseed rape (RO), rice–green manure (RG) (in 2018, rice–hairy vetch (RH)), and rice–winter fallow (RF) (Figure 7.2). According to the local straw return mode, the rice field was plowed first and then rotated during the rice season, and the wheat, rape, and green manure were all adopted less or no tillage measures were undertaken. Each treatment was repeated three times for a total of 12 plots with a plot area of 60 m^2, and the fertilizer and water management measures were in accordance with the local high-yield cultivation technology mode. Different experiments of the new mode for carbon sequestration and emission reduction showed that compared with the rice–wheat mode (RW), other modes had the effect of reducing emissions from the perspective of annual greenhouse gas emissions. It was worth noting that among the winter crops, the treatment of green manure had the largest CH_4 emissions, and the difference was significant. However, the CH_4 emissions of the entire system were mainly concentrated in summer, so the trend of CH_4 emissions for the entire production system remained the same as that of the summer rice in the system. Rice–wheat (RW) had the highest CH_4 emissions, and rice–winter fallow treatment (RF) had the lowest CH_4 emissions. Among different planting patterns, rice–wheat (RW) always had the highest N_2O emissions rate, followed by rice–oilseed rape (RO), rice–green manure (RG), and rice–winter fallow (RF). Considering the effects of emission reduction,

Figure 7.2.   Screening and demonstration of new modes for carbon sequestration and emission reduction in wheat–rice project area.

rice yield, winter crop biomass, and other factors, the rice–green manure (RG) planting mode is a more suitable choice.

(1) The impact of different crop rotation modes on greenhouse gas emissions

The results in 2017 showed that under different crop rotation modes, the change in the trend of CH_4 emission fluxes from winter crops was not obvious. Before mid-February, the CH_4 emission rate of planting modes of rice–wheat (RW) and rice–green manure (RG) was low, and the emission of rice–oilseed rape (RO) and rice–winter fallow (RF) planting modes was relatively high. After mid-February, there were emission peaks in each rotation planting system, but the time was different. There was no similar rule between treatments, and the difference between them was not significant. The results in 2018 showed that the CH_4 emission rate of winter planting of green manure treatments rice–oilseed rape (RO), rice–hairy vetch (RH)) before April 11 was higher than that of rice–wheat (RW) and rice–winter fallow (RF). However, CH_4 emission increased after the winter planting of green manure, which might be caused by the larger biomass of green manure crops forming shelter (such as vetch), which caused the increase of soil moisture. The experimental results in 2018 showed that there were significant differences in CH_4 emission rates in rice seasons under different planting patterns. There were two CH_4 emission peaks during the entire rice-growing season. The first one occurred on July 11, and the CH_4 emission rates of rice–wheat (RW), rice–oilseed rape (RO), rice–hairy vetch (RH) and rice–winter fallow (RF) planting mode were 32.6, 23.1, 18.7 and 4.6 mg CO_2-eq m^{-2} h^{-1}, respectively. Among them, the rice–wheat mode had the highest emission rate and the rice–winter fallow (RF) mode had the lowest emission rate. The second one occurred between August 9 and August 30, and the rates of each treatment for the second peak of emissions were 12.5, 14.9, 12.2, and 22.5 mg CO_2-eq m^{-2} h^{-1} respectively.

The experimental results in 2017 showed that the N_2O emission rate trends of winter crops under different crop rotation patterns were not significant, but overall, there were two emission peaks: The first time was in mid-February, and this was because the temperature rose with the return of spring. The N_2O emission rate gradually increased as the temperature rose and reached the peak. The second peak was in mid-April, which was because of the dual effects of late topdressing and temperature rise. Among them, the rice–wheat (RW) and the rice–oilseed rape (RO) modes had similar trends, and the N_2O emission rate was relatively high; the rice–wheat (RW) system had the largest N_2O emissions, the rice–green manure (RG) and rice–winter fallow (RF) planting mode had lower N_2O emissions than the other two modes, and the rice–winter fallow (RF) planting mode had the lowest N_2O emission rate. The experimental results in 2018 showed that except for the winter crops in the rice–hairy vetch (RH) treatment, which had an emission peak (316.1 μg CO_2-eq m^{-2} h^{-1}) on April 18, the N_2O emission peaks of other treatments all appeared on March 30. During this period, the N_2O emission rates of winter crops under the rice–wheat (RW), rice–oilseed rape (RO), rice–hairy vetch (RH), and rice–winter fallow (RF) modes were 393.5, 121.1, 103.9, and 87.8 μg CO_2-eq m^{-2} h^{-1}, respectively. The 2-year experimental results showed that the rice–wheat (RW) mode had the highest emission rate, and the rice–hairy vetch (RH) mode had the lowest emission; the rice–winter fallow (RF) planting mode was an effective mode for reducing N_2O emissions from winter crops. There was no significant difference in N_2O emissions between the treatments in the rice season under different planting patterns. In mid-to-late July, each treatment had an emission peak. Among them, the N_2O emission of the rice–hairy vetch (RH) and rice–winter fallow (RF) planting modes in the rice season was relatively high, reaching 220.3 and 204.0 μg CO_2-eq m^{-2} h^{-1}. This was because the paddy field was vacant in August, and the paddy soil was well aerated, which promoted the activity of microorganisms and accelerated N_2O emissions.

(i) Cumulative greenhouse gas emissions during the growth period of the rotation mode. Comparing different crop rotation modes, there were significant differences in greenhouse gas emissions: The rice–wheat (RW) planting mode had the highest N_2O emission, reaching 1,261.1 kg CO_2-eq hm^{-2}, followed by the rice–oilseed rape (RO) mode and the rice–green manure (RG) mode, and the rice–winter fallow (RF) mode had the lowest emissions. Compared with the rice–wheat mode, the N_2O emissions under the rice–oil, rice–milk vetch, and rice–winter fallow rotation modes were reduced by 7.4%, 33.9%, and 70.5%, respectively. Among the four crop rotation modes, both the modes of planting rape and milk vetch in winter had lower CH_4 emissions. After further comparing the greenhouse gas emissions per unit area (GWP) under different crop rotation modes, it was found that both planting rape and milk vetch in winter were beneficial to reducing the GWP, but the GWP of the winter fallow mode was the lowest.

The study results showed that CH_4 emissions from the rice field rotation system were mainly concentrated in the rice-growing season. Compared with the rice–wheat rotation mode (RW), CH_4 emissions of the rice–oilseed rape (RO), rice–hairy vetch (RH) (green manure), and rice–winter fallow (RF) rotation modes in the rice season was reduced, and the rice–winter fallow (RF) rotation mode had the largest decline. Among the winter crops, rice–hairy vetch (RH) had the largest CH_4 emission, which was significantly different from other treatments. The N_2O emissions from the paddy field rotation system were mainly concentrated in the winter crop-growing season, and there was no significant difference between the treatments. The greenhouse gas emissions per unit area data showed that compared to the rice–wheat (RW) rotation system, the greenhouse gas emissions per unit area of other crop rotation systems had all decreased, and the GWP of the rice–winter fallow (RF) planting mode was the smallest, so winter fallow could effectively reduce greenhouse gas emissions.

Analyzing the annual CH_4 emissions of crops under different crop rotation modes, it could be seen that CH_4 emissions were mainly concentrated in the rice season, and the CH_4 emissions of winter crops were very small. In 2016, the rice–wheat (RW) rotation mode had the largest CH_4 emission in the rice season and the lowest CH_4 emission obtained from the rice–winter fallow (RF) mode. Because the CH_4 emissions in the rice season accounted for a relatively large proportion, the trend of annual CH_4 emissions for each treatment was consistent with the rice season. Compared with the rice–winter fallow (RF) rotation mode, the annual CH_4 emissions of rice–wheat (RW), rice–oilseed rape (RO), and rice–green manure (RG) in 2016 increased by 69.7%, 40.6%, and 25.5%, respectively. In 2017, the rice–wheat (RW) mode decreased by 22.1%, while the rice–oilseed rape (RO) and rice–green manure (RG) modes increased by 32.3% and 24.6%, respectively. This showed that using green manure as a winter crop could effectively reduce CH_4 emissions in the rice season.

Compared with the rice–winter fallow (RF) rotation mode, the annual N_2O emissions of rice–wheat (RW), rice–oilseed rape (RO), and rice–green manure (RG) in 2016 increased by 113.6%, 11.4% and 8.4%, respectively; in 2017, the rice–wheat (RW), rice–oilseed rape (RO), and rice–green manure (RG) modes decreased by 28.0%, 38.8% and 3.6%, respectively.

The results of annual crop global warming potential of different planting patterns showed that greenhouse gas emissions in the rice season had a greater impact on the annual global warming potential; the rice–wheat (RW) rotation mode had the largest global warming potential, and the rice–winter fallow (RF) rotation mode had the lowest global warming potential; in comparison, the annual GWP emissions of rice–wheat (RW), rice–oilseed rape (RO), and rice–green manure (RG) in 2016 increased by 55.2%, 28.8%, and 20.1% respectively; in 2017, only the rice–wheat (RW) mode decreased by 21.1%, the rice–oilseed rape (RO) and rice–green manure (RG) modes increased by 33.1% and 24.6%, respectively.

(2) The impact of different crop rotation modes on output

Comparing the effects of different crop rotation modes on rice yield, it was found that under the condition of returning all winter crop straw to the field, compared with the rice–winter fallow (RF) rotation mode, rice yields of the rice–wheat (RW), rice–oilseed rape (RO), and rice–green manure (RG) modes in 2016 increased by approximately 4.8%, 5.6%, and 11%, respectively. In the winter of 2016, the crops encountered cold damage after sowing, which severely reduced the emergence rate of wheat, rape, and milk vetch; therefore, the output of the rice season in 2017 was affected to a certain extent. Under different crop rotation modes, the rice–wheat (RW) rotation mode had the lowest rice yield of 9.83 t hm^{-2}, and the rice yields of the rice–oilseed rape (RO), rice–hairy vetch (RH), and rice–winter fallow (RF) modes were, respectively, 10.54, 10.63, and 10.57 t hm^{-2}, which increased by 7.8%, 8.5%, and 7.8%, respectively. Among them, the green manure and vetch had the best effect on increasing the rice yield. Among the biomass of winter crops, the biomass of hairy vetch was the largest, reaching 5.29 t hm^{-2}, followed by wheat. Therefore, from the perspective of rice yield and winter crop biomass, using rice–vetch (RG) crop rotation has the highest economic benefits, and it could increase the yield significantly.

7.1.3 *Demonstration of Supporting Cultivation Techniques for Conservation Tillage*

The tillage treatment of rice season in 2015 was divided into four groups: (i) rotary tillage in wheat season+plow tillage in rice season+full straw return to the field (RT-CT-S), (ii) no-tillage in wheat season+rotary tillage in rice season+full straw return to the field (NT-RT-S), (iii) rotary tillage in wheat season+plow and rotary tillage in rice season (RT-CT), and (iv) no-tillage in wheat season+rotary tillage in rice season (NT-RT) (Figure 7.3).

The wheat seasons in 2016 and 2017 consisted of eight tillage treatments: (i) plow and rotary tillage in rice season+rotary tillage in wheat season+full straw return to the field (CT-RT-S), (ii) plow and rotary tillage in rice season+no-tillage in wheat+full straw return to the field (CT-NT-S), (iii) rotary tillage in rice season+rotary tillage in wheat season+full straw return to the field (RT-RT-S), (iv) rotary tillage in rice season+no-tillage in wheat season+full straw return to the field (RT-NT-S), (v) plow and rotary tillage in rice season+rotary tillage in wheat season (CT-RT), (vi) plow and rotary tillage in rice season+no-tillage in wheat season (CT-NT), (vii) rotary tillage in rice season+rotary tillage in wheat season (RT-RT), and (viii) rotary tillage in rice season+no-tillage in wheat season (RT-NT). The experimental results of supporting cultivation techniques of different conservation tillage from 2015 to 2018 showed that the full straw returned to the field could reduce N_2O and CH_4 emissions in the wheat and rice seasons. Under the condition of returning the full straw to the field, because of the treatment of rotary tillage in the wheat season, the effect of greenhouse gas emission reduction was more significant. The treatment of plow and rotary tillage in the rice season could reduce the N_2O emissions of the two crops in the entire rice–wheat production system. Under the condition of full straw returning to the field, the treatment of rotary tillage in wheat season was beneficial to increasing and stabilizing the yield, but there was no significant difference in the effect of the rice yield with plow and rotary tillage in the rice season.

(1) Impacts of different farming methods and straw returning methods on wheat–rice greenhouse gas emissions
CH_4 and N_2O emission dynamics in the rice field. In 2015, after comparing the CH_4 emission rate of rice fields during the rice growth period with experiments, it was found that the overall change was unimodal. There were significant differences between the treatments from July 7 to August 2, and the emission peak was reached on July 14. At the peak of emission, the emission rate of the treatment of rotary tillage in wheat season+plow and rotary tillage in rice season (RT-CT) was the largest, and no-tillage in wheat season+rotary tillage in rice season+full straw return to the field (NT-RT-S) was the smallest, and the full straw return to the field could reduce the peak emission rate of CH_4 emissions.

Figure 7.3.   Demonstration of supporting cultivation techniques for conservation tillage in the wheat–rice project area.

The experimental results in 2015 showed that the N_2O emission rate from rice fields under different tillage methods showed a multi-peak trend. The first peak was on July 14, no-tillage in wheat season+rotary tillage in rice season+full straw return to the field (NT-RTS), and no-tillage in wheat season+rotary tillage in rice season (NT-RT) had higher emission rates, which were significantly higher than those in the treatments of rotary tillage in wheat season+plow and rotary tillage in rice season (RT-CT) and rotary tillage in wheat season+plow and rotary tillage in rice season+full straw return to the field (RT-CT-S), indicating that rotary tillage in wheat season+plow and rotary tillage in rice season (RT-CT) reduced the first emissions of N_2O. The second peak was on August 20; this time the N_2O emissions under the treatments of rotary tillage in wheat season+plow and rotary tillage in rice season (RT-CT) and no-tillage in wheat season+rotary tillage in rice season (NT-RT) were higher than the treatments of rotary tillage in wheat season+plow and rotary tillage in rice season+full straw return to the field (RT-CT-S) and no-tillage in wheat season+rotary tillage in rice season+full straw return to the field (NT-RT-S). The third time was on September 22 when this emission peak showed that the N_2O emission of the treatment under rotary tillage in wheat season+plow and rotary tillage in rice season (RT-CT) was significantly higher than the other three treatments. The experiment results showed that the N_2O emission rate of the treatment without straw returning to the field was higher than that of the treatments with straw returning to the field.

(i) CH_4 and N_2O emission dynamics in wheat field. The CH_4 emissions in the wheat season were relatively small. The experimental results in 2016 and 2017 showed that the CH_4 emissions of the treatment without straw returning was greater than that of the treatment with full straw returning, and there was no significant difference in CH_4 emissions between different treatments. The experimental results in 2016 showed that two N_2O emission peaks occurred during the entire growth period of wheat under different tillage treatments. The first emission peak occurred in mid-February when the N_2O emission rate under the treatment without straw returning to the field reached an average of 143.2 μg CO_2-eq m^{-2} h^{-1} and the N_2O emission rate under the treatment with full straw return to the field decreased by about 23.8%. The N_2O emission rate of the rotary tillage treatment was slightly lower than that of the no-tillage treatment. The second emission peak appeared in mid-to-late March, and the trend was similar to the first emission peak. In 2017, the experimental results showed that N_2O emissions during the whole wheat growth period had two peaks under different tillage treatments. The first one occurred in mid-to-late February, and the N_2O emission rate under the treatment without straw returning to the field reached an average of 380.0 μg CO_2-eq m^{-2} h^{-1}, and the N_2O emission rate under the treatment with full straw return to the field increased by about 21.8%. The N_2O emission rate of the rotary tillage treatment was slightly lower than that of the no-tillage treatment. The second one occurred in mid-April because of the dual effects of top dressing and warming. Rotary tillage in wheat season+no-tillage in rice season+full straw return to the field (RT-NT-S) had the highest emission rate, followed by rotary tillage in rice season+rotary tillage in wheat season (RT-RT). These two treatments were higher than other treatments. Comparing the N_2O emission rate during the whole growth period of the wheat season, the N_2O emission rate of the rotary tillage treatment in the rice season was higher than that of the rotary tillage treatment in the rice season. Therefore, the plow and rotary tillage treatment in the rice season was beneficial to reducing the system N_2O emissions.

(ii) Annual greenhouse gas emissions from rice–wheat rotation. Further analysis of the effects of different tillage methods and straw returning methods on the greenhouse effect in rice fields showed that CH_4 emissions of the plow and rotary tillage treatment in rice season was reduced, especially under the condition of all straw returning to the field, with the reduction rate was 11.7%. Regardless of whether the straw was returned to the field or not, the N_2O emissions of the plow and rotary tillage measure in rice season was reduced; after the full amount of straw was returned to the field, the N_2O emission from the rice field was significantly reduced. Compared with no straw returning to the field, the N_2O emission was reduced by about 26.3%. The comprehensive greenhouse effect and emission intensity change trend of rice fields were consistent with CH_4 emissions.

Soil farming methods have a significant impact on greenhouse gas emissions in the wheat season. The experimental results in 2016 showed that the average N_2O and CH_4 emissions under the rotary tillage treatment in the wheat season reached 1161.7 and 46.2 kg CO_2-eq hm^{-2}, respectively. Compared with the less or not tillage treatment, the N_2O emissions reduced by about 20.2%, and CH_4 increased by about 624.3%. Compared with the straw returning to the field, the average N_2O and CH_4 emissions under the condition of not returning the straw to the field reached 1,062.0 and 3.8 kg CO_2-eq hm^{-2}, respectively. Compared with not returning straw to the field, their emissions reduced by approximately 31.7% and 88.5%, respectively. A further comparison of GWP and emissions per unit yield (GHGI) in the wheat season found that although the rotary tillage in wheat season significantly increased CH_4 emissions, its contribution to the GWP was relatively small, so compared with the less or not tillage treatment, the GWP and GHGI of rotary tillage in wheat season reduced by about 16.6% and 16.2%, respectively. Compared with not returning straw to the field, the GWP and GHGI of the wheat season under the full amount of straw returned to the field reduced by about 32.9% and 28%, respectively. In other words, under the condition of returning all the straws in the wheat season to the field, combined with rotary tillage measures, the emission reduction effect was better.

The experiments in 2017 showed that the average N_2O and CH_4 emissions under the rotary tillage treatment in wheat season reached 8,023.2 and 96.2 kg CO_2-eq hm^{-2}, respectively. Compared with the less or not tillage treatment, the N_2O emissions increased by about 20.2%, and CH_4 emissions increased by about 624.3%; for the straw returning to the field, the average N_2O and CH_4 emissions under the condition of not returning the straw to the field reached 7,514.3 and 102.6 kg CO_2-eq hm^{-2}, respectively. Compared with no returning, the N_2O emissions of the full amount of straw returned to the field reduced by approximately 8.7% and 26.7%, respectively. A further comparison of GWP in the wheat season found that although the rotary tillage in wheat season significantly increased N_2O and CH_4 emissions, compared with the wheat season, the GWP of rotary tillage in wheat season increased by about 20.8%. Compared with not returning straw to the field, the GWP of the wheat season under the full amount of straw returned to the field reduced by about 9%.

A study showed that the global warming potential of the wheat season was reduced by about 10.1% after the full amount of straw was returned to the field, but the rice season increased by about 3.5%. Compared with the non-return of straw, the average global warming potential of the 2 years after the full annual straw was returned to the field had little change (0.6% increase). A further comparison of the differences in different tillage methods found that compared with the rotary tillage treatment, the 2-year average GWP of the no-tillage treatment in the wheat season increased by 17.0%, but the plow tillage treatment in the rice season decreased by 6.9%.

(2) Effects of different tillage methods and straw returning methods on yield

The experiments in 2015 showed that the treatment of straw returning to the field had little effect on rice yield. The rice yields under the treatments of rotary tillage in wheat season+plow and rotary tillage in rice season+full straw return to the field (RT-CT-S), no-tillage in wheat season+rotary tillage in rice season+full straw return to the field (NT-RT-S), rotary tillage in wheat season+plow and rotary tillage (RT-CT), and no-tillage in wheat season+rotary tillage in rice season (NT-RT) were 13.29, 13.75, 13.45 and 13.53 t hm^{-2}, respectively. Compared with not returning the straw to the field, the rice yield under the full straw return to the field increased by about 2%. The rice yield under the rotary tillage treatment was slightly higher than that under the plow tillage treatment; compared with the plow tillage treatment, the rice yield under the rotary tillage treatment increased by about 2.1%, but the difference between the treatments was not significant. In the 2016 wheat season, wheat yield under the treatments of rotary tillage in wheat season+full straw return to the field (RT-S), no-tillage in wheat season+full straw return to the field (NT-S), rotary tillage in wheat season (RT), and no-tillage in wheat season (NT) were 5.70, 5.48, 6.23 and 6.65 t hm^{-2} respectively. Compared with returning straw to the field, the yield of wheat without straw returned to the field increased by 12.9%. Under the condition of not returning the straw to the field, compared with the rotary tillage measures, the wheat yield under the condition of less or not tillage measure increased by about 6.95%, but the difference did not reach a significant level. The result was the opposite when the full straw was returned to the field, and the wheat yield under the less or no tillage measure decreased by about 3.54%. It can be seen that under the condition of the full amount of straw was returned to the field, an appropriate increase in tillage intensity, such as rotary tillage, would help stabilize the yield.

The experiment of tillage measures in 2016 and 2017 showed that after the rice straw was fully returned to the field, the changes in wheat yield under different tillage methods were different. For the wheat seasons in 2016 and 2017, rotary tillage treatments were larger than the no-tillage treatments; compared with the no-tillage treatments, the wheat yield under rotary tillage measures increased by 6.1%, and the trends were the same in 2 years. Under the condition that the straw was fully returned to the field in the rice season, compared with the rotary tillage treatment, the rice yield of the plow tillage treatment increased by about 8.4% in 2016, while it decreased by about 7% in 2017, but the difference between the treatments for 2 years did not reach a significant level. In terms of annual output, compared with the rotary tillage treatment, the 2-year average annual crop yield of the plow tillage treatment under the condition that the straw was fully returned to the field was reduced by about 1.7%, and the difference between treatments was not significant. The experimental results showed that the treatment of rotary tillage in the wheat season was beneficial to increasing the yield of wheat, while the plow and rotary tillage measures in the rice season had no significant effect on the yield of rice.

7.2 Training and Consulting Service of the Climate-smart Wheat–Rice Production Technology

The project in Huaiyuan area focused on the concepts, policies, and technologies related to CSA and invited leaders from the Ministry of Agriculture and Rural Affairs, directors from the Department of Agriculture and Rural Affairs of Anhui Province, experts in related fields such as China Agricultural University, Chinese Academy of Agricultural Sciences, Anhui Agricultural University, and local leaders in Huaiyuan and others giving lectures for local farmers, agricultural machinery personnel, and village cadres. It conducted hands-on on-site teaching of agricultural techniques that farmers were concerned about in the rice–wheat production system, such as conventional agricultural operations and pest control, and carried out a series of field classroom activities. In response to the national rural revitalization strategy, a new combination of planting and breeding technology was introduced locally, and the supporting agricultural product marketing strategies were explained and trained to farmers, so that farmers could not only plant and raise well, but also had better sales, truly improving the agricultural yield of local farmers and increasing farmers' income. This also helped to further enhance the understanding of the theories and technologies of climate-smart agriculture by local village cadres, agricultural technicians, and large grain farmers, and effectively promoted the taking roots of climate-smart agriculture. During the period 2016–2019, the project team compiled the *Concept and Mode of Climate Intelligent Agriculture Development, High-Quality and High-Yield Cultivation Techniques of Mid-Japonica Rice, Integrated Control Techniques of Wheat Diseases, Pests and Diseases, Wheat Soil Testing and Formula Fertilization and Nitrogen Fertilizer Backwashing Technology, Technical Specifications for Mechanized Operation of Wheat Straw Returning to Field under Rice–wheat Double-cropping,* and *Wanfu Town Spring Wheat Field Management Technology* in 2017, *the key rice-duck technology,* and other training materials, and conducted more than 20 agricultural technology training classes in the Huaiyuan project area, organized and trained more than 2,000 people including agricultural technicians, large grain farmers, and village and town cadres. These efforts achieved good results and effectively promoted the dissemination of CSA concepts and technologies.

7.2.1 *Training on the Concept and Practice of CSA*

In order for local village officials, agricultural technicians, large growers, and ordinary farmers to have a deeper understanding of the theory and practical effects of CSA, the project training team invited well-known experts and organized more than 10 special lectures on "Climate-Smart Agriculture", including "Technical Training of Climate-Smart Major Food Crop Production Projects" and "Theory and Practice of Energy Saving and Emission Reduction of Farmland Ecosystem", which systematically explained the principles, main technologies, and production applications of CSA, effectively improved farmers' acceptance of new technologies and new production mode of fit. In addition, the training experts effectively combined the "The leader who became wealthy after starting own business in poor villages in Huaiyuan County" with CSA, and also how the leader who became wealthy after starting own business in poor villages adapted to climate change, they had carried out structural adjustments and industrial upgrades, and systematically explained and trained them to "contribute to poverty alleviation".

7.2.2　Technical Training and Guidance of Rice–Wheat Production Management

The Huaiyuan project area is a high-quality rice–wheat production area in our country, and the optimizing management measures such as water and fertilizer under rice–wheat high-yield conditions are an important way to improve crop production efficiency and reduce greenhouse gas emissions (Figure 7.4). From 2016 to 2019, the Huaiyuan project area training team organized more than 20 special training reports including "High-yield cultivation and fertilizer-water management technology of rice", "Mechanized production technology and equipment special training of rice", "High-efficiency planting mode optimization and precise fertilization technology", "Rice light and simplified cultivation technology", "Rice fertilizer–water coupling technology and rice mid-late management", "Spring management and jointing fertilization technology of wheat under rice stubble", "Precision balanced fertilization technology for rice–wheat dual-cropping", "Precision fertilization technology for rice", "Soil ecological restoration and wheat fertilizer-water management technology", "Law of rice–wheat fertilizer requirement and precision fertilization technology", "Planting and feeding combination mode of green, low-carbon and cyclic and its key technologies", "Discussion on green, high-quality and high-efficiency cultivation techniques of wheat under rice stubble", "Application technology of jointing fertilizer for wheat under rice stubble", "Cultivation technology and field management of wheat", "High-quality production and food security of wheat", and "Precision fertilization technology of wheat".

The training reports includes not only the basic characteristics of high-yield rice, the cultivation technology of strong seedlings

Figure 7.4.   Classroom training in wheat–rice project area.

by machine-transplanted rice, the precise quantitative fertilizer-water management technology for the field of machine-transplanted rice, the new problems brought about by the coupling technology of fertilizer-water for rice, and the innovation of the middle-late rice management farming system. It also includes field water management, weed control, jointing fertilizer application, and other issues of wheat under rice stubble in spring. At the same time, these reports showed a deep analysis on how to choose wheat varieties under climate change conditions, and how wheat cultivation techniques adapt to climate change. Taking the continuous rain in the autumn and winter of 2018–2019, which caused the late sowing of wheat under the rice stubble and weakened the seedlings as an example, detailed explanations were given on how to do a good job in spring field management, especially the scientific application of stand-up jointing fertilizer. The training reports also emphasized that rice–wheat production should pay attention to the green and low-carbon production mode, scientific fertilizer-water management, established and improved soil environmental quality testing, and provided a healthy soil environment for ecological agriculture, thereby developing green and pollution-free organic agricultural products; the reports also systematically introduced the green and low-carbon cycle development concept, the necessity and possibility of implementing the combination of planting and breeding, demonstrated the successful mode of the combination of planting and breeding in Anhui, and through examples in detail introduced the main modes and key technologies of the combination of planting and breeding. After 4 years of consecutive classroom training, the local agricultural machinery personnel, large grain growers, and farmers have been greatly improved on the theory and technology of high-yield and efficient cultivation and management of water, nitrogen, and wheat, which had effectively promoted the application and popularization of climate-smart agricultural technology in demonstration areas.

7.2.3 *Rice–Wheat Straw Returning Technology*

The amount of rice and wheat straw is large. Rice–wheat straw returning to the field is an important way to improve farmland soil organic matter content and reduce straw pollution, and the straw returning technology is an important part of climate-smart agricultural technology. From 2016 to 2019, the Huaiyuan project area training team invited experts in related fields to organize more than 10 special training reports including the "Green yield increasing mode and straw returning mechanization technology", "Rice mechanized straw returning technology", "Agricultural machinery and agronomy technology for rice–wheat production", "Combination of agricultural machinery and agronomy under the conditions of straw returning to the field", "Mechanized straw returning to the field and green high-yield cultivation technology of wheat", "Mechanized returning to the field and comprehensive utilization technology of crop straw", "Mechanized technology of returning straw to the field for rice under wheat stubble", "Mechanized straw returning to field of rice stubble and site preparation technology of wheat", and "Equipment and key technology of straw returning to field in rice–wheat cropping area", and "Mechanized straw returning technology".

These reports focused on the combination of green and high-yield cultivation techniques, such as agricultural machinery and agronomy technology of returning straw to the field, precise formula fertilization technology, fine soil preparation, selective breeding, proper sowing, reasonable dense planting, uniform sowing, diseases, and pest prevention. In particular, detailed explanations were given on the use of chemical

Figure 7.5.   Field technical guidance for wheat–rice project area.

fertilizers, the planting density of wheat, and the prevention and treatment of frost damage in late spring. The detailed explanations focused on the comprehensive utilization of crop straw, the mechanized return of straw to the field in wheat–maize rotation, the mechanized return of straw to the field in the rice–wheat rotation, the mechanized collection of straw, the introduction of wheat straw in the field, the selection of different agricultural machinery and the corresponding technical points, the technical measures of rice–wheat double-cropping system in the technical measures of mechanized straw returning to the field, especially the key technology of the mechanized returning of wheat straw to the field. The technical measures should achieve "six degrees": cutting stubble height, straw cutting length, spreading uniformity, tillage depth, irrigation depth, and nitrogen fertilizer increase range. It put forward requirements for crushing wheat straw: straw crushing length (less than 10 cm), cutting stubble height (less than 15 cm), and straw laying (evenly spreading and avoiding striped laying). Choose a combine harvester with sufficient power to install the straw crushing device, and fast disassembly should be the functional goal. The technical key points of straw returning technology of rice–wheat double cropping system and new methods of mechanized production of wheat under rice stubble were mainly explained. These had effectively improved the understanding of local agricultural technicians, large growers, and farmers on the straw returning technology for rice–wheat annual production and effectively promoted the comprehensive popularization of the straw returning technology for rice–wheat annual production in the Huaiyuan project area (Figure 7.5).

7.2.4 *Integrated Green Control of Diseases and Pests in Rice-Wheat System*

Diseases and pests were important factors restricting the high yield of rice and wheat in the Huaiyuan project area (Figure 7.6). The project training team organized well-known experts in related fields to carry out more than 10 special training reports including "Technology for the prevention and control of wheat diseases and pests under rice stubble", "Green prevention and control technology for rice diseases and pests", "Research progress and green prevention and control of rice and wheat diseases", "Pesticide residues and agricultural product quality safety", "Technology for preventing and controlling rice diseases and pests in late stage", " Main diseases and insect pests of rice and wheat and their control", "Integrated control technology for wheat diseases and insect pests under rice stubble", "Comprehensive control technology for diseases, pests and weeds in rice–wheat double cropping system", "Green control technology for rice diseases and pests", "Integrated control technology of wheat diseases and pests", and "Green prevention and control of rice and wheat pests".

Figure 7.6.   Disease and pest control in the wheat–rice project area.

From the aspects of overall prevention and control technology for rice and wheat, experts professionally gave introduction and training on integrated prevention and control of diseases and insect pests of single-season rice, and integrated prevention and control methods of main diseases and insect pests of double-season late rice. Prevention and control of "three insects and three diseases" that rice most often faces are as follows: (1) three insects: rice planthopper, rice leaf roller, and rice stem borer (borer) and (2) three diseases: sheath blight, rice blast, false smut, and major wheat diseases (sheath blight, rust disease, powdery mildew, and gibberellic disease). The three major wheat pests are sitodiplosis mosellana, aphid, and wheat spiders. Combining the photo examples, experts also gave introduction on the occurrence laws of diseases and pests of wheat and rice under the rice stubble in spring, and the onset period, location, and symptoms of diseases and pests. The green control technology of diseases, pests, and weeds was explained, so timely chemical control and environmental protection were achieved. Through training, awareness of local agricultural technicians and major grain farmers on the occurrence and prevention of diseases and pests in rice–wheat production had been greatly improved, and the adverse effects of diseases and pests on rice–wheat production had been effectively reduced.

7.2.5 *New Farming Technology*

Increasing crop yield and farmer's income is an important part of climate-smart agriculture, and the development of new planting and breeding technologies is an important way to increase farmers' income (Figure 7.7). The project training team organized more than 10 special reports including the "Modern agriculture and agricultural structure adjustment", "Modern agriculture and supply-side agricultural structure adjustment", "Interpretation and development trend of domestic and foreign rice-fishing integrated planting and breeding status", "General situation of aquatic vegetable specialty areas in our country and pollution-free lotus root cultivation techniques", "Rice-lobster ecological comprehensive planting and breeding technology", "Green prevention and control technology of pests in rice–shrimp field", "Fertilizer-water management technology", and "Comprehensive planting and breeding". The reports focused on how to integrate the three industries, how to adjust the agricultural planting structure, variety structure, planting and breeding structure, and exploring new modes of high-yield, high-efficiency, energy-saving, and emission reduction under CSA. The reports took rice–crayfish ecological farming as an example and gave detailed introduction from the perspective of crayfish farming, the general situation of the development of crayfish farming, and understanding the planting and breeding mode of rice, the comprehensive planting and breeding in rice field that were mainly promoted in Anhui Province, problems in lobster breeding, and selection of lobster breeding species from a new height. Taking Quanjiao County's rice–crayfish ecological planting and breeding technology mode as an example, reports comprehensively analyzed the key technologies of rice–crayfish ecological comprehensive planting and breeding. After the lecture, through interactions, experts answered technical questions about the green prevention and control of rice diseases and insect pests and fertilizer-water management in shrimp fields. At the same time, other issues were also explained, such as the continuous cropping of rice and shrimp introduced in detail in terms of the construction of rice field breeding projects, the selection and stocking of crayfish fry, the feeding of feed, water quality management, daily management, rice cultivation and management, and the catching and harvesting of crayfish.

Figure 7.7.   New planting and breeding technologies in wheat–rice project area.

7.2.6 *Comprehensive Field Technical Guidance for Crop Production*

The project team organized crop production experts, plant protection, and agricultural product quality and safety experts to open up field classrooms to provide technical consultation, guidance, and technical services with large growers and farmers on rice–wheat high-yield variety selection, optimal management of water and fertilizers, key technologies, and management methods of three-dimensional planting and breeding modes such as paddy field duck and rice field fish breeding. The core technical issues of pesticide production also were given attention, such as integrated prevention and control of rice–wheat pests and diseases, unmanned aerial vehicle fertilization and field operation of spraying pesticides, and rice–wheat straw returning to the field. Experts helped to solve farmer's crop production problems through on-site diagnosis and discussion, such as prevention and control of diseases and pests and straw returning technology. Besides, experts deepened farmer's application of drones in fertilizer and pesticide application through on-site display and explanation (Figure 7.8).

Figure 7.8.   Field comprehensive technical guidance for wheat–rice project area.

7.2.7 *Marketing of Agricultural Products*

The sale of agricultural products is an important section in agricultural production, especially for large farmers with plenty of fields and products. Enhancing farmers' marketing awareness is conducive to increasing farmers' income. The project training team organized more than 10 special reports, such as the "Basic theory and strategy of modern product marketing", "Food quality and safety", "Modern marketing strategies for agricultural products", "Theory and strategy of modern agricultural product marketing", and "Strategy and case analysis of upward operation of agricultural product".

Reports from the No. 1 Central Document and the government work reports the development of e-commerce, from sales of electricity to the transformation of industrial electricity, through fusion of online and offline, combination of far and near services, and agricultural travel experience cracked the market competition homogeneity bottleneck, let students understood the plight of the current rural electricity; through vivid examples, the reports emphatically analyzed how to sell agricultural products online, not only to strengthen the construction of brand, attach great importance to the safety of agricultural products, electric business platform, and the supply chain, but also to emphasize the key of talent training. Taking Shanggu company as an example in Anhui, reports focused on "the dual-drive mode of brand and trading platform boosting Huaiyuan's agricultural upgrade".

The training reports also gave a detailed and vivid description of the basic theories of agricultural product marketing and the practical application of marketing strategies. We not only need to produce good agricultural products, but also should pay more attention to the deep processing and marketing of agricultural products, so as to increase farmers' income and build a new socialist countryside. Reports also emphasized the development of high-yield, high-quality, high-efficiency, ecological, and safe modern agriculture, improving the level of the quality and safety of agricultural products and also enhancing the competitiveness of agricultural products. The quality and safety of agricultural products should be paid attention to by the farmers, and there was an explanation in detail about the current situation of agricultural product quality and safety, quality safety certification of "three products and one standard", and organic agricultural product processing certification. The importance of brand trademarks was explained in detail. Brand trademarks are an important part of the overall product and an important intangible asset of an enterprise.

7.2.8 *Theory and Practice of Rural Revitalization*

Rural revitalization is a direct manifestation of rural social development promoted by CSA, and the project training team organized thematic training reports, including the "Beautiful countryside construction and comprehensive environmental improvement", "Rural revitalization strategy and rural development", "Decisive battle against poverty", and "Building three modern agriculture systems to promote industrial precision alleviation" (Figure 7.9). Reports systematically introducing the 2017 Central Committee No. 1 Document *Several Opinions of the Central Committee of the Communist Party of China and the State Council on Deepening the Advancement of Agricultural Supply-side Structural Reform and Accelerating the Cultivation of New Driving Forces for Agricultural and Rural Development*, pointed out that the project area should seize the opportunity of structural reform on the agricultural supply-side, while ensuring grain output, improving grain quality, adjusting the structure of agricultural production, integrating the development of primary, secondary, and tertiary industries and exploring new models of high-yield, high-efficiency, energy-saving, and emission reduction. Combined with the current actual situation

Figure 7.9. Rural revitalization in wheat–rice project area.

of Huaiyuan County, it was proposed that Huaiyuan County's implementation of the rural revitalization strategy should strictly implement the cultivated land protection system and comprehensively improve the quality of cultivated land, optimize the agricultural industry structure and actively carry out large-scale operations, improve water efficiency and ensure safe agricultural water use, adjust and optimize livestock-poultry breeding layout and rationally set up the scale of livestock-poultry breeding, strengthen the construction of livestock-poultry manure treatment facilities, improve the level of recycling of agricultural resources, vigorously promote formula fertilization and fully implement green prevention and control technologies, and comprehensively promote the construction of beautiful villages and accelerate improvement in rural ecological environment.

Visiting and studying in the national agricultural park. The project team organized agricultural technicians, and large grain growers and farmers to visit and study in Shandong Shouguang International Vegetable Science and Technology Expo, Lujiang County National Modern Agriculture Demonstration Zone, Anhui Youxin Agricultural

Figure 7.10.   Visiting and learning of national agricultural park of wheat–rice project area.

Technology Company, Anhui Area of Shanghai Green Super Rice Research and Development Center, Anhui Nongkang Farm, and the Wanzhong Comprehensive Experimental Station of Anhui Agricultural University and other places (Figure 7.10). They learned about the progress and achievements in agricultural science and technology, and experienced and learned about the new concepts, new achievements, new equipment, new materials, new varieties, and new management methods of modern agriculture. Everyone earnestly learned the cultivation of vegetable seedlings, agricultural products and vegetable processing, the biological control of vegetable diseases and insect pests, the application of high-tech, the technical links of factory seedling raising, and how to deal with adverse climate effects, the status quo of three-dimensional cultivation of rice and shrimp, breeding technology, and breeding risk and avoidance technologies. This had greatly stimulated the participants' enthusiasm for learning and agricultural production enthusiasm, and strongly promoted the application of climate-smart agricultural technology in the project area.

7.3 The Monitoring and Evaluation of Climate-smart Wheat–Rice Production

Establishing an appropriate dynamic monitoring and evaluation mechanism is an important way for timely feedback and change of project implementation strategies and tracking the effect of the project (Figure 7.11). By comparing with the baseline, timely and accurately assessing the progress of CSA projects, reflecting on the problems existing in the project implementation, and proposing corresponding countermeasures, the mechanism could effectively ensure the progress of the project and achieve the set goals. During the project period from 2016 to 2019, we focused on tracking the crop production, carbon sequestration, and emission reduction, environmental effects, pest management, and effects and social impact in the project area. The yield of rice and wheat in the project area was higher than that under the baseline conditions. The project conditions could effectively reduce greenhouse gas emissions during the rice season, increase soil organic carbon storage, forest carbon sinks, and improve surface water, farmland retreat, and groundwater quality. The disease and insect pests, pesticide use, and medical equipment in the project area had all improved compared with the baseline conditions.

7.3.1 *The Monitoring and Evaluation of Carbon Sequestration and Emission Reduction*

For the climate-smart wheat–rice cropping system, the project team provided scientific support for CSA development based on these aspects: determining the project area, project boundary, project target group, and project activity objects; carrying out the project area monitoring and evaluation of plan design, and selecting the sample villages, the sample farmers and the sample farmland, set up monitoring points and monitoring areas in a targeted manner, based on the application of the relevant implementation technology of the project; carrying out continuous and dynamic monitoring of relevant control indicators to assess greenhouse gas emissions and soil carbon storage in the climate-smart agricultural production system in the project area. We studied and predicted the overall carbon sequestration and emission reduction capabilities of the project area and optimized the improvement approach through field observations and the DNDC model to quantify the carbon sequestration and emission reduction potential of climate-smart agriculture.

Figure 7.11.   Winter wheat–rice planting pattern monitoring in the wheat–rice project area.

Since 2015, monitoring points have been arranged in Huaiyuan County, Anhui Province, and experimental monitoring was initiated. According to the requirements of the project and the needs of technical research, we systematically monitored the changes in soil carbon sinks, greenhouse gas emissions, yield composition, and so on, aiming to test the carbon sink emission reduction effects of the project.

(1) Dynamic monitoring of crop yield

During the implementation of the project, the project team selected 120 yield monitoring samples from wheat–rice system in Huaiyuan County, Anhui Province, for the yield information collection, of which five replicate samples were taken from each sampling point (Figure 7.12).

The monitoring results showed that there were certain differences in winter wheat yield in different project areas. In 2016, the winter wheat yield under the project conditions in the project area in Huaiyuan County, Anhui Province was far lower than the winter wheat yield under the baseline conditions. In 2016, under the baseline conditions in the Huaiyuan project area, the winter wheat yield was 6,410 kg ha^{-1}, and the yield under the project conditions was 5,723 kg ha^{-1}, in which the project conditions was 10.7% lower than the baseline conditions. In 2017, the winter wheat yield under the project conditions was greater than the project yield under the baseline conditions. In 2017, in the project area of Huaiyuan County, Anhui Province, the winter wheat yield under the project conditions was 7,928 kg ha^{-1}, and the yield under the baseline conditions was 7,630 kg ha^{-1}, with the project conditions being 5.3% higher than the baseline conditions. In 2018, the winter wheat yield under the project conditions in the Anhui project area showed an increasing trend compared to the baseline

Figure 7.12.   Production site of the wheat–rice project area.

conditions. The winter wheat yield under the project conditions was 6,851 kg ha^{-1}, which was 5.1% higher than the baseline conditions. The decrease in winter wheat yield under the project conditions in the first year compared with the baseline conditions was due to the greater disturbance of the soil and crop growth caused by agricultural management measures, which would have a greater impact or even decrease in yield. However, after the second year of project implementation, due to the reduction of interference from various measures on winter wheat, the yield of winter wheat under project conditions had increased, and the project's yield increase or stabilization effect began to appear, achieving the expected results.

The results of rice monitoring in the project area of Huaiyuan County, Anhui Province, showed that from 2016 to 2018, the rice yields under the project conditions were higher than those under the baseline conditions. In 2016, the rice yield under the project conditions was 8,303 kg ha^{-1}, and the yield under the baseline conditions was 8,170 kg ha^{-1}, with the output increased by 1.6% under project conditions. Field monitoring was conducted after harvesting in October 2017, in which the average of different monitoring points yield was measured. The rice yield under the baseline conditions in the Huaiyuan project area in Anhui was 7,928 kg ha^{-1}, and the yield under the project conditions was 8,605 kg ha^{-1}, which was an 8.5% increase under the baseline yield. In 2018, the rice yield under the project conditions was 9,129 kg ha^{-1}, which was an increase of 5.9% compared to the baseline of 8,622 kg ha^{-1}. Judging from the implementation of the 3 years, the implementation of the project has a satisfactory effect on rice production.

(2) The characteristics of greenhouse gas emissions from farmland
(i) N_2O emissions during the wheat season. The N_2O emissions of winter wheat in the Huaiyuan County, Anhui Province, in 2016 and 2017 were different. In 2016, the N_2O emissions of winter wheat under the project conditions were greater than that under the baseline conditions, while in 2017, the emissions under the project conditions were less than the baseline conditions. In 2016, in the project area of Huaiyuan, the N_2O emissions during the growth period of winter wheat under the baseline conditions was 3.47 kg N ha^{-1}, while the emissions under the project conditions was 4.19 kg N ha^{-1}, the emissions under project conditions increased by 20.75% compared to baseline conditions. In 2017, in the Huaiyuan project area, the N_2O emissions of winter wheat under the baseline and project conditions were 11.92 and 9.29 kg N ha^{-1}, respectively,

and the N_2O emissions under the project conditions were reduced by 22.06% compared with the baseline conditions.

(ii) CH_4 and N_2O emissions during the rice season. The rice season emits two greenhouse gases, CH_4 and N_2O. In the first 2 years of project implementation, the N_2O emissions under project conditions were higher than baseline conditions. In 2018, the N_2O emissions under project conditions were lower than under baseline conditions. During the 3-year project implementation period, the CH_4 emissions under the project conditions were all less under than the baseline conditions, and the reduction rates were 9.64%, 5.56%, and 4.07%. CH_4 is the main greenhouse gas. The test results showed that the CH_4 emissions of rice during the project implementation were reduced, which led to the lowering of greenhouse gas emissions during the rice season.

(3) Soil carbon storage

The soil carbon storage in the project area is calculated according to the calculation method in *CSA-C-3-Carbon Sink emissions Reduction Monitoring and Evaluation*, in which the results showed that during 2016–2018, the soil carbon storage under the technical conditions of the two project areas increased, that is, intervention measures under project conditions could enhance the carbon sink function of the soil. It also showed that as the time of project intervention increases, it would enhance the soil's carbon sink function.

The 2019 carbon sink reports showed that the organic matter contents of the top 0–30 cm soil layer in project area varies from 19.51 to 20.59 g kg^{-1}; the average soil organic carbon content per unit area was (61.82 $\pm$ 0.94) t C ha^{-1}, which was 14.79% higher than the baseline. The demonstration area of carbon sequestration and emission reduction technology in Huaiyuan is 2,000 hectares. Together with 2018 reports, we could conclude that the 2019 soil carbon storage of top 0–30 cm soil layer in the project area was increased by 16,904 t CO_2 eq compared to 2018 (Table 7.1).

(4) Forest carbon sink

From 2016 to 2018, under project conditions, the afforestation area and forest carbon sink in the project area continued to increase (Figure 7.13). In 2016, the number of trees planted in the Huaiyuan project area was

Table 7.1.   Soil carbon storage changes in Huaiyuan project area from 2016 to 2019.

Year	Condition	DSOC (t C ha⁻¹)	Area (ha)	ΔDSOC (t C ha⁻¹)	ΔSOC (t CO₂)
	Baseline	53.86 ± 0.62			
2016	Project	54.26 ± 0.784	33.33	0.40	643
2017	Project	56.19 ± 3.09	966.67	1.93	6,839
2018	Project	59.52 ± 0.78	1, 766.67	3.33	21,544
2019	Project	61.82 ± 0.94	2,000.00	2.311	6,904

Figure 7.13.   Forest perimeter detection in the wheat–rice project area.

calculated at 7,000 trees per year, and the amount of carbon sequestered was 143.94 t C. In 2017, the Huaiyuan project area completed more than 66.66 ha of afforestation and planted more than 20,000 seedlings of various types such as paulownia. According to the formula calculation, the forest carbon sink of Huaiyuan County reached 1,118.02 t C yr⁻¹. In 2018, the Huaiyuan project area completed more than 66.66 ha of afforestation, planted 20,000 paulownia and other seedlings, and the forest carbon sink reached 1,672.03 t C yr⁻¹.

7.3.2 *Environmental Effect Monitoring and Evaluation*

As the international communities pay more and more attention to climate change, greenhouse gas emissions reduction and food security, the research on farmland soil carbon sequestration and emission reduction technology have received unprecedented attention from the scientific community (Figure 7.14). Through promoting the application of energy-saving

Figure 7.14.   Environmental effect detection of wheat–rice project area.

and carbon sequestration technologies under the premise of ensuring food production in major grain production areas, and conducting demonstration and evaluation of emission reduction effects, we can improve soil fertility and productivity and reduce soil greenhouse gas emissions. It is a strategic choice for our country to maintain sustainable agricultural development.

The area of rice–wheat rotation pattern in Huaiyuan County is large enough to represent the development status of China's paddy and dry rotation areas. The local agricultural cooperation organization is completed, and the application of new agricultural machinery and agricultural technology is well established. The rice–wheat rotation system demonstration zone in the project area was composed of 12 administrative villages, including the union of Wanfu Town and Lanqiao Town, Sunzhuang Village, Zhendong Village, Qiannan Village, Liulou Village, Chen'an Village, Zhuanqiao Village, Houji Village, Guanxu Village, Huihan Village, Yuxia Village, and Zhaomu Village.

Since 2015, various programs have been conducted in the project, including soil testing and formula fertilization, formula deep fertilization, crop reduction application, straw return, and land leveling. These reduced the application of chemical fertilizers and pesticides and reduced the return of farmland water. It also reduced the agricultural pollution sources entering the waters of the project area and reduced the risk of harmful substances and nutritional pollutants entering the water body through the monitoring of surface water and groundwater in the project area, thus the environmental benefits were determined.

(1) Arrangement of monitoring points in the project area
The main planting mode in the project area of Anhui Province is rice–wheat rotation. The irrigation depends on water from Cihuaixin River and Qianhe River, where farmland irrigation water is discharged into Qinhe River and farmers' use of shallow groundwater as drinking water. Soil types can be divided into two categories: umber soil and sand ginger black soil. Sampling in the Huaiyuan project area includes the following: (i) river irrigation water — the sampling site was located at the water intake section of the Cihuaixin River project area and the water intake section of the Qianhe River intake project area; (ii) water effluent in the project area — the sampling point was located at the water outlet section of the Qianhe project area; (iii) 20 meters of shallow groundwater — selected from Lianhe Village, Qianmu Village, Liulou Village, Zhuanqiao Village, and Zhennan Village (Table 7.2). The survey and sampling determined the location of the sampling points by considering the soil map and administrative division map of the project area and the irrigation form of the project area and formed the location map of the sampling points.

(2) Sampling and testing contents
According to the information on irrigation type, soil classification, crop planting area, and river distribution in the project area, the surface water would be monitored at least twice a year and the measurement indicators included CODCr, $NH+_4$-N, TN, TP, and NO^-_3-N. Groundwater is the source of drinking water for residents in the project area, and the quality of groundwater is very important to the health of local people. According to the irrigation type, soil classification, and planting area of the project area, the groundwater would be monitored at least twice a year and the measurement indicators included permanganate index, NO^-_3-N, TN, TP, and $NH+_4$-N. The main evaluating methods of water quality monitoring include the single-factor evaluation method, multi-factor evaluation method, and entropy weight fuzzy matter-element evaluation model. In this study, the entropy-weighted fuzzy matter-element model method was selected to calculate the weight in the multi-factor comprehensive evaluation. The entropy weighted

fuzzy matter-element model has a great resolution for each evaluation index, and its weight value will play a decisive role in the evaluation result and is a key factor of evaluation. The weight adopts the entropy method, and its basic principle is based on the degree of change of each index value. The greater the degree of index value changes, the smaller the information entropy value will be, indicating that the greater the amount of information provided by the index, the greater the corresponding weight is; on the contrary, the smaller the weight will be. This method can avoid the error caused by human scoring and use the change of the factor itself to objectively obtain the weight value for evaluation.

(3) Water quality in the project area
The water quality monitoring of project area started in 2014 and was conducted in January 2014, June 2015, May 2016, December 2016, April 2017, July 2017, and December 2017. Sampling and testing were carried out in April 2018, November 2018, May 2019, and December 2019. A total of five monitoring points were selected for monitoring in Huaiyuan County. The monitored indicators included the chemical oxygen demand (COD), ammonium content ($NH+_4$-N), total phosphorus (TP), and total nitrogen (TN) content of the water body. The monitored waterbodies included surface water and groundwater. The monitoring results of surface water were evaluated according to the national standard (GB 3838–2002), and the groundwater was evaluated according to the national standard (GB/T 14848–93).

(i) Surface water monitoring results.
The analysis of the monitoring results of surface water showed that from 2014 to 2019, because there were fewer agricultural activities throughout the winter, and almost no activities such as fertilizer application and irrigation, the water quality in summer was worse than in winter. In the spring, as the frequency increases in agricultural activities such as fertilization, irrigation, spraying pesticides, and drainage, the number of water pollutants discharged from farmland were also increasing, and the water quality would deteriorate. Among the surface water monitoring indicators, the main changing indicators were COD and TN.

For the surface water observation points in various places, the major change in water quality was COD, and this change will be larger with the increase of the project implementation period. The COD content in the water body showed a downward trend. It showed that the implementation of the project had a certain positive effect on reducing the COD content in water bodies. As far as water quality of the surface water observation points in various places was concerned, with the increase of the project implementation period, the water quality was showing an improvement trend. Although the quality of water body is improving, the TN content in the water body showed a slight increasing trend in most surface water observation points. It showed that the implementation of the project might increase the risk of TN loss.

Table 7.2.   2014–2019 Cihuaixin River inlet water quality monitoring results (mg l⁻¹).

Site	Sample time	COD	NH_4^+-N	TP	TN	Water category	Quality
Intake of Cihuaixin River	January 2014	8.49	0.1	0.06	1.15	IV	
	June 2015	15.4	0.08	0.03	1.26	IV	
	May 2016	<15	0.17	0.01	0.605	III	
	December 2016	11.9	0.15	0.05	1.89	V	
	April 2017	6.9	0.08	0.02	0.899	III	
Intake of Cihuaixin River	July 2017	14.5	0.2	0.07	1.18	IV	
	December 2017	<2.3	0.21	0.08	0.575	III	
	April 2018	6	0.2	0.02	1.1	IV	
	November 2018	5.7	0.1	0.1	1.21	IV	
	May 2019	4.0	0.30	0.04	0.61	III	
	December 2019	12.9	0.111	0.03	1.00	III	
Intake of Ci River	January 2014	20.91	0.1	0.96	0.06	IV	
	June 2015	27.9	0.33	0.04	0.74	IV	
	May 2016	30	0.42	0.05	0.78	IV	
	December 2016	24.6	1.51	0.06	9.8	Inferior V	
	April 2017	16.6	0.37	0.11	0.691	III	
	July 2017	64	1.06	0.19	1.49	Inferior V	
	December 2017	22.1	0.28	0.08	0.689	IV	
	April 2018	21	0.48	0.62	0.95	Inferior V	
	November 2018	8	0.28	0.09	0.67	III	
	May 2019	16.0	0.34	0.16	0.62	III	
	December 2019	18.0	0.214	0.04	0.56	III	

(Continued)

Table 7.2.   (*Continued*)

Site	Sample time	COD	NH$_4^+$-N	TP	TN	Water category	Quality
Zhennan Village	June 2015	133	1.71	0.93	3.3	Inferior V	
	April 2017	6.6	0.24	0.04	1.25	IV	
	July 2017	30	0.72	0.29	1.06	IV	
	December 2017	18.2	0.52	0.33	2.21	Inferior V	
	April 2018	12	0.36	0.06	1.17	IV	
	November 2018	4.6	0.11	0.09	0.43	II	
	May 2019	10.0	0.53	0.32	1.07	V	
Outtake of Ci River	January 2014	22.57	0.14	1.6	0.13	V	
	June 2015	25.8	0.34	0.09	1.15	IV	
	May 2016	25	0.38	0.08	0.698	IV	
	December 2016	21.4	1.19	0.07	5.79	Inferior V	
	April 2017	14	0.22	0.06	0.581	III	
	July 2017	27.2	0.46	0.19	1.11	IV	
	December 2017	11.3	0.1	0.07	0.565	III	
	April 2018	13	0.27	0.07	0.86	III	
	November 2018	5	0.38	0.13	0.91	III	
	May 2019	13.0	0.34	0.32	0.77	V	
	December 2019	17.5	0.173	0.03	0.59	III	
Zhaomu Village	June 2015	27.7	0.33	0.05	0.71	IV	
	May 2016	29	0.37	0.12	0.997	IV	
	December 2016	27.2	1.74	0.23	2.73	Inferior V	
	April 2017	21.3	0.75	0.11	1.08	IV	
	July 2017	55	5.4	1.11	7.33	Inferior V	
	December 2017	40	0.38	0.07	0.998	V	
Zhaomu Village	April 2018	45	1.39	0.41	2.66	Inferior V	
	November 2018	2.3	0.34	0.26	0.71	III	
	May 2019	16.0	1.07	0.26	1.35	IV	

(ii) Observations of farmland receding water. Sampling observations were also completed during the project implementation of the farmland retreat in Huaiyuan Country, Anhui Province. The observation results are shown in Table 7.3: The quality of water retreats in different seasons of the same year had certain differences, and the quality in summer was worse than that in winter. During the project implementation period, the

Table 7.3.   2015–2018 Huaiyuan County farmland receding water quality monitoring results (mg l⁻¹).

Sample time	CODCr	NH_4^+-N	TP	TN	Water quality category
January 2015	22.43	2.09	0.17	7.92	Inferior V
June 2015	62	1.69	0.5	2.88	Inferior V
June 2015	29.7	0.49	0.18	1.55	V
June 2015	29.4	0.47	0.21	1.42	IV
September 2015	<15	0.58	0.17	1.17	IV
September 2015	27	0.62	0.16	1.38	IV
December 2015	22	0.19	0.04	0.234	IV
December 2015	<15	0.28	0.06	0.306	II
May 2016	227	6.62	3.49	6.82	Inferior V
June 2016	30	1.08	0.47	1.67	Inferior V
July 2016	30.8	3.37	0.32	6.34	Inferior V
July 2016	34.3	0.73	0.36	2.18	Inferior V
August 2016	45	0.75	0.25	2.38	Inferior V
August 2016	33.5	0.45	0.2	1.5	V
August 2016	34.5	0.53	0.18	1.63	V
August 2016	23.8	0.3	0.14	1.09	IV
December 2016	13.9	0.38	0.05	1.1	IV
December 2016	8.5	0.09	0.05	0.515	III
April 2017	9.8	0.1	0.1	0.472	II
July 2017	53	2.14	0.29	2.51	Inferior V
December 2017	11.8	0.15	0.08	0.72	III
May 2018	19	0.35	0.08	0.91	III
December 2018	2.3	0.14	0.1	0.35	II
December 2019	18	0.617	0.41	1.82	Inferior V
December 2019	6.3	0.098	0.04	0.85	III

quality of farmland receding water showed a gradual improvement trend with the implementation of the project. Especially after 2018, the quality of farmland receding water Class III in summer and Class II in winter. Compared with the data in 2015, there had been a significant improvement. It showed that the implementation of the project was conducive to improving the quality of farmland receding water.

Among the measurement indicators of farmland receding water, COD had a large range of changes and showed a downward trend in fluctuations. In the same month of different years, as the years of the project progressed, the COD of farmland receding water was decreasing. In June 2015, the COD content of farmland receding water was 62 mg l^{-1} in June 2016, the COD content of farmland receding water was reduced to 30 mg l^{-1}; the COD content of 2015, 2016, 2017, and December 2018 was 22, 13.9, 11.8, and 2.3 mg l^{-1}, and the downward trend was very obvious. In addition to the COD, the variation of TN in farmland return water was also very obvious. During the implementation of the project, the TN content in farmland receding water had also been continuously fluctuating but decreasing. It was also shown that the relevant measures taken in the project could effectively reduce the loss of TN in farmland, and played an important role in reducing the loss of farmland nutrients and water pollution.

(iii) Groundwater observation results. The groundwater monitoring results in Huaiyuan County, Anhui Province, were shown in the Table 7.4. Since June 2015, the quality of groundwater in the project area was between Class I and II. In January 2014, before the project was implemented, the quality of groundwater at the monitoring point was poorer than that after the implementation of the project. Thus, the implementation of the project could improve the quality of groundwater in the project area.

From the analysis of surface water, groundwater, and farmland water retreat in project area, the implementation of the project could reduce the content of pollutants in the water body and improve the quality of water body.

7.3.3 *Pest Management Monitoring and Evaluation*

According to the actual agricultural production in the project area, during the implementation of the project, the use of pesticides and fertilizers would be reduced under the precondition of ensuring food production, comprehensive prevention and control of pests, control of pests and diseases, and reduction of pesticide pollution. Thus, the implementation of the project should pay more attention to the application of integrated pest

Table 7.4.   2014–2019 Groundwater monitoring results in the project area of Huaiyuan County, Anhui Province (mg l^1).

Sample time	Site	Permanganate index	TP	NH^+_4-N	NO^-_3-N	Water quality category
January 2014	Liulou Village	2.24	0.01	0.11		III
	Zhuanqiao Village	1.94	0.07	0.09		II
	Zhennan Village	3.02	0.03	0.05		IV
	Lianhe Village	2.21	0.03	0.07		III
	Zhaomu Village	1.81	0.02	0.11	0.143	II
June 2015	Liulou Village	0.73	0.01	0.22	0.37	I
	Zhuanqiao Village	0.89	0.01	0.3	0.38	I
	Lianhe Village	0.62	0.02	0.05	0.37	I
May 2016	Lianhe Village	0.58	0.09	<0.050	0.384	I
	Zhennan Village	1.08	0.09	<0.050	0.373	II
	Liulou Village	0.92	0.06	0.08	0.344	I
	Zhuanqiao Village	1	0.07	<0.050	0.387	I
May 2017	Lianhe Village	0.4	0.06	0.451	0.272	I
	Liulou Village	0.48	0.06	0.524	0.268	I
	Zhennan Village	0.55	0.05	0.42	0.272	I
	Zhuanqiao Village	0.55	0.06	0.545	0.269	I
July 2017	Lianhe Village	0.53	0.1	0.519	0.294	I
	Liulou Village	0.72	0.11	0.8	0.324	I
	Zhennan Village	0.49	0.08	0.654	0.324	I
	Zhuanqiao Village	0.52	0.1	0.795	0.262	I
May 2018	Lianhe Village	0.65	0.09	0.44	0.4	I
	Liulou Village	0.69	0.08	0.46	0.38	I
	Zhennan Village	0.89	0.08	0.46	0.4	I
	Zhuanqiao Village	0.65	0.08	0.42	0.41	I
December 2018	Lianhe Village	1.25	0.1	0.15	0.15	II
	Liulou Village	1.7	0.08	0.32	0.15	II
	Zhennan Village	1.49	0.07	0.38	0.37	II
	Zhuanqiao Village	1.49	0.1	0.4	0.4	II
May 2019	Lianhe Village	0.74	0.04	<0.05	<0.15	I
	Liulou Village	0.65	0.06	0.3	<0.15	I

(*Continued*)

Table 7.4. (*Continued*)

Sample time	Site	Permanganate index	TP	NH_4^+-N	NO_3^--N	Water quality category
	Zhennan Village	0.48	0.03	0.16	<0.15	I
	Zhuanqiao Village	0.48	0.06	0.13	<0.15	I
December 2019	Lianhe Village	0.4	0.04	<0.05	<0.009	I
	Liulou Village	0.32	0.03	0.16	<0.009	I
	Zhennan Village	0.32	0.03	<0.05	<0.009	I
	Zhuanqiao Village	0.32	0.03	0.22	<0.009	I

management technology and specialized in unified prevention and control. Integrated pest management is a strategy of Integrated Pest Management (IPM). From the perspective of the agricultural ecosystem, and based on the relationship between pests and the environment, it coordinated various measures such as agricultural, physical, biological, and chemical prevention and control, and exerted the role of natural control factors in agricultural ecology, making agriculture biological control to be below the allowable level of economic losses. IPM attaches great importance to the application of integrated control technologies including resistant varieties, cultivation measures, biological natural enemies, and chemical agents, especially the use of biological control factors such as natural enemies to control pests and diseases and adopts a cautious attitude towards the use of chemical pesticides. The specialized prevention and control of crop diseases and insect pests refers to the service organizations with corresponding plant protection expertise and equipment to carry out socialized, large-scale, and intensive crop diseases and insect pest prevention and control services. Specialized control is the product of agricultural development to a certain stage, which conforms to the general law of agricultural development in the world today and is an important support for the implementation of the concept of "public plant protection and green plant protection". It is also an important measure to promote the stable growth of food production and various cash crops, an effective means to ensure the quantity and quality safety of agricultural products and the safety of the agricultural ecological environment, and an important guarantee for agricultural efficiency, farmers' income, and rural stability.

Figure 7.15.   Detection and evaluation of pests and diseases in Huaiyuan project area.

(1) The technical plan of farmer's production survey

The occurrence of pests and crop yield surveys. The project team visited Huaiyuan County, Anhui Province, in September 2015, September 2016, December 2016, April 2017, July 2017, June 2018, August 2018, June 2019, and August 2019. Nine farmer surveys were carried out to assess the occurrence of wheat and rice pests during the implementation of the project (Figure 7.15).

In the investigation of the occurrence of rice diseases and insect pests, in 2015, eight rice farmers were selected randomly from the demonstration villages in Huaiyuan County, Anhui Province, and six rice farmers were selected randomly from non-demonstration villages to test the project's impact on the occurrence of rice diseases and insect pests. In 2016, one rice grower (large household) was selected from the demonstration village, and five rice growers were selected randomly from non-demonstration villages to test the impact of the project on rice yield. In 2017, 2018, and 2019, a large rice farmer was selected randomly from a demonstration village, and nine rice growers were selected randomly from non-demonstration villages to test the impact of the project on the occurrence of rice diseases and insect pests.

In the investigation of the occurrence of wheat diseases and insect pests, based on 27 wheat farming households selected randomly from the demonstration villages in 2015 (including eight households in the wheat–rice cropping system and 19 households in the wheat–maize cropping system) and nine wheat farmers (including six households in the wheat–rice cropping system and three households in the wheat–maize cropping system) selected randomly from non-demonstration villages, the impact of the project on the occurrence of wheat diseases and insect pests was assessed. In 2016, based on one wheat farmer selected from the demonstration villages of the wheat–rice cropping system, and five wheat farmers selected from the non-demonstration villages to test the impact of the project on wheat production. In 2017, 2018, and 2019, based on eight wheat farming households selected from the demonstration villages (including one household in the wheat–rice cropping system and seven households in the wheat–maize cropping system), and 10 wheat farming households (including nine households in the wheat–rice cropping system and one household in the wheat–maize cropping system) selected from non-demonstration villages, the impact of the project on the occurrence of wheat diseases and insect pests was assessed.

(i) Investigation of pest control. The project team carried out nine surveys to track the use of pesticides by farmers in September 2015, September 2016, December 2016, April 2017, July 2017, June 2018, August 2018, June 2019, and August 2019 in the project area of Huaiyuan County in Anhui Province. The survey subjects selected the same diseases and insect pests. The survey investigated the number of times, types of pesticides used, and the total amount of pesticides used by farmers in the demonstration and non-demonstration villages in 2015 and 2016, and calculated the cost of pesticides in the production costs of various crops. The survey subjects selected the same disease and insect pests' occurrence survey to test the impact of the implementation of the project on the yield of wheat and rice crops. The survey subjects selected the same diseases and insect pests to investigate the situation, and the use of pesticides and equipment in the demonstration villages and non-demonstration villages. In the survey, the number of pesticides, the types of pesticides, the amount of pesticides per 667 m^2, and the total amount of pesticides were investigated and compared, and the cost of pesticides in the production costs of various crops was calculated.

The large applicators and unmanned applicators purchased during the project implementation in 2015 and 2016 had not been used on a large scale due to various reasons. The most used by farmers was the conventional sprayer for spraying. In 2017, 2018, and 2019, the demonstration area of Huaiyuan County, Anhui Province used large-scale wide-width pesticide applicators purchased by large growers during the implementation of the project. Farmers in non-demonstration areas used ordinary knapsack electric sprayers. In 2017 and 2018, a pesticide use test was conducted in Zhuanqiao Village, Wanfu Town, Huaiyuan County, Anhui Province. The Allure Red indicator was used as the experimental indicator, and the static electricity was measured by referring to the process of the determination of the effective use rate of pesticides by the Institute of Plant Protection, Chinese Academy of Agricultural Sciences. The experiment measured the effective use of pesticides in electrostatic sprayers, large sprayers, and ordinary electric sprayers.

(2) The results of the investigation of crop diseases and insect pests
(i) Diseases, pests, and weeds in rice field. In 2015, the average occurrence number of pests and diseases among rice farmers in the demonstration villages were 6.75, including three times of diseases, 2.875 times of pests, and 0.875 times of weeds; the average occurrence number of diseases and insects in non-demonstration villages was 7.167, of which 3.167 times of diseases, three times of pests and one time of weeds. The average occurrence number of diseases, pests, and weeds in non-demonstration villages was higher than that in demonstration villages, but the difference was not significant. In 2016, the average occurrence number of pests and diseases of rice growers in the demonstration villages were six times, including three times of diseases, three times of pests, and 0 time of weeds. The average number of pests and diseases in non-demonstration villages was six including three times of diseases, three times of pests, and 0 time of weeds. The average occurrence number of rice diseases, pests, and weeds in the demonstration villages and non-demonstration villages were the same. In 2017, the average occurrence number of pests and diseases of rice farmers in the demonstration villages were seven times, including three times of diseases, two times of insect pests, and two times of weeds; the average occurrence number of pests and diseases of non-demonstration villages were seven times, including three times of diseases, two times of pests, and two times of weeds. The average occurrence number of rice diseases, pests, and weeds in the demonstration villages and non-demonstration villages were the same. In 2018, the average occurrence of the number of pests and diseases of rice farmers in

demonstration villages were seven times, including three times of diseases, two times of insect pests, and two times of weeds; the average occurrence number of pests and diseases of non-demonstration villages were 7.1, including three times of diseases, two times of pests and 2.1 times of weeds. There was no significant difference in the average number of rice diseases, pests, and weeds between the demonstration villages and non-demonstration villages. In 2019, the average occurrence number of pests and diseases of rice growers in the demonstration villages were six times, including two times of diseases, two times of pests, and two times of weeds. The average number of pests and diseases in non-demonstration village were 7.1, including three times of diseases, two times of pests, and 2.1 times of weeds. It could be seen that the average occurrence of rice pests and diseases in the demonstration villages had decreased compared with non-demonstration villages, indicating that the prevention and control of rice pests and diseases in the demonstration villages was better.

(ii) Diseases, pests, and weeds in the wheat field. In 2015, the average occurrence number of pests and diseases among wheat growers in demonstration villages were 5.778, of which 1.926 were diseases, 2.074 were pests, and 1.778 were weeds; the average occurrence number of pests and diseases in non-demonstration villages were 6.889, including three times of diseases, 2.333 times of pests, and 1.556 times of weeds. The average occurrence number of wheat diseases and insect pests in non-demonstration villages was higher than that in the demonstration villages. In 2016, the average occurrence number of pests and diseases of wheat growers in demonstration villages was eight, including three times of diseases, two times of pests, and three times of weeds; the average occurrence number of pests and diseases in non-demonstration villages were nine, including two times of diseases, four times of pests, and three times of weeds. The average occurrence number of wheat diseases and insect pests in non-demonstration villages was more than that in the demonstration villages. Among them, the average occurrence number of diseases was one less than that in the demonstration villages. The occurrence number of weeds were the same, but the average occurrence of pests in non-demonstration villages was two times more than that in the demonstration villages. In 2017, the average occurrence number of pests and diseases of wheat growers in the demonstration villages were five, including two times of diseases, two times of pests, and one time of weeds; the average occurrence number of pests and diseases in non-demonstration villages were seven, including three times of diseases, three times of pests, and one time of weeds. The average occurrence number of wheat diseases and insect pests in non-demonstration villages were more than that in demonstration villages, the average occurrence of diseases and insect pests was one more than in the demonstration villages, and the occurrence number of weeds were the same. In 2018, the average number of pests and diseases of wheat growers in the demonstration villages were five, including two times of diseases, two times of pests, and one time of weeds; the average occurrence number of pests and diseases in non-demonstration villages were 7.1, including 3.1 times of diseases, three times of pests, and one time of weeds. The average occurrence number of wheat diseases and insect pests in non-demonstration villages were more than that in the demonstration villages, the average occurrence number of diseases and insect pests was one more than in the demonstration villages, and the occurrence of weeds were the same. In 2019, the average occurrence number of pests and diseases of wheat growers in the demonstration villages were five, including two times of diseases, two times of pests, and one time of weeds; the average occurrence number of pests and diseases in non-demonstration villages were 7.9, including 3.3 times of diseases, 3.3 times of pests, and 1.3 times of weeds. It could be seen that the average occurrence number of wheat diseases, pests, and weeds in non-demonstration villages were more than that in demonstration villages. Among them, the average frequency of disease and insect pests were basically one to two times more than that of the demonstration village, and the average occurrence number of weeds was almost one time higher than that of the demonstration village, indicating that the demonstration village had better practices in the prevention and control of wheat diseases and insect pests, and the demonstration effect was better.

(3) The survey results of pesticide use
(i) The use of pesticide in rice cultivation. In 2015, the average application number of medications in rice demonstration villages and non-demonstration villages were 3.13 and three times respectively, and the difference were not obvious. The average cost of pesticides per 667 m^2 in demonstration villages was ¥9.68, and that in non-demonstration villages was ¥88.95. The cost of medication per 667 m^2 in demonstration villages was much lower than that in non-demonstration villages. In 2015, the average number of pests, diseases, and weeds among rice growers in the demonstration villages were lower than that in non-demonstration villages. Under the guidance of project experts, farmers in the demonstration

villages increased their use of pesticides, and the yield per 667 m^2 increased by 40% compared with non-demonstration villages. The implementation effect was remarkable. In 2016, the average number of medications in rice demonstration villages and non-demonstration villages were four and 3.1 times, respectively. The average number of pesticides in demonstration villages was one time more than that in non-demonstration villages. The average cost of pesticides per 667 m^2 in the demonstration villages was ¥100, and that in non-demonstration villages was ¥72.14. The cost of medication per 667 m^2 in demonstration villages was ¥27.86 higher than that in non-demonstration villages. In 2016, the average number of pests and diseases among rice growers in the demonstration villages was the same as that in the non-demonstration villages. The demonstration villages used pesticides once more than the non-demonstration villages, but the yield of the demonstration villages was 16% lower than that of the non-demonstration villages. The average yield per 667 m^2 in 2016 in the demonstration villages was 13% lower than that in 2015, and the yield per 667 m^2 in non-demonstration villages in 2016 was 48% higher than in 2015. It was estimated that factors other than pesticides (droughts, floods, and other meteorological disasters) in 2016 would have a greater impact on rice production in demonstration and non-demonstration villages. In 2017, the average application number of pesticides in rice demonstration villages and non-demonstration villages were six and seven times, respectively, and the difference was the same. The average cost of pesticides per 667 m^2 in demonstration villages was ¥65.5, and the average cost of pesticides per 667 m^2 in non-demonstration villages was ¥90.0. The cost of pesticides per 667 m^2 in demonstration villages was much lower than in non-demonstration villages. In 2017, the average number of pests and diseases among rice growers in demonstration villages was lower than that in non-demonstration villages. Under the guidance of project experts, the average yield per 667 m^2 increased by 10% compared with non-demonstration villages, and the project implementation effect was remarkable. In 2018 and 2019, the average number of medications in rice demonstration villages and non-demonstration villages was six and seven, respectively, and the difference was the same. The average cost of medicine per 667 m^2 in demonstration villages was ¥63, and the average cost of medicine per 667 m^2 in non-demonstration villages was ¥92. In 2018, the average number of pests and diseases among rice growers in demonstration villages was lower than that in non-demonstration villages. Under the guidance of project experts, the average yield per 667 m^2 increased by 10% compared with non-demonstration villages, and the project implementation effect was remarkable. Among the farmers in the demonstration village, there was a large planting household with a planting area of 600 ha, of which 400 ha of self-sown rice fields and 200 ha of machine-transplanted rice fields were much higher than other farmers' planting scales. Thanks to the preferential price of pesticides purchased in large quantities and the high use rate of pesticides applied on a wide range, the cost of pesticides was reduced.

(ii) The use of pesticide in wheat cultivation. In 2015, the average number of times of pesticide use in wheat demonstration villages and non-demonstration villages were 2.8 and 3.7 times, respectively, and the average number of times in non-demonstration villages were 0.9 times more than that in the demonstration villages. The average cost of pesticides per 667 m^2 in demonstration villages was ¥42.0, and the average cost of medication per 667 m^2 in non-demonstration villages was ¥64.2. The cost of pesticides per 667 m^2 in the demonstration villages was ¥22.2 lower than that in non-demonstration villages. In 2015, the number of occurrences of diseases, pests, and weeds and the number of pesticide applications in the demonstration villages were lower than those in the non-demonstration villages, but the average yield per 667 m^2 was 20% lower. Besides diseases, pests, and weeds, other factors had a greater impact on yield. In 2016, the average use times of wheat in demonstration villages and non-demonstration villages were 4 and 2.5 times, respectively, and the average use times in demonstration villages was 1.5 times more than that in non-demonstration villages. The cost of pesticides in demonstration villages was provided by the government, and the average cost of pesticides per 667 m^2 in non-demonstration villages was ¥42.9. In 2016, the number of wheat diseases, pests, and weeds in the demonstration villages was lower than that in the non-demonstration villages, the number of pesticides were 1.5 times higher, and the yield per 667 m^2 was 16% higher. The project implementation effect was remarkable. In 2017, the average number of times of pesticides use in wheat demonstration villages and non-demonstration villages were three and four times, respectively, and the average number of times of pesticides use in non-demonstration villages was one time more than that in demonstration villages. The average cost of pesticides per 667 m^2 in demonstration villages was ¥23.6, and that in non-demonstration villages was ¥34.5 per 667 m^2. The cost of pesticides per 667 m^2 in the demonstration villages was ¥10.9 lower than that in the non-demonstration villages. In 2017, the number of occurrences of diseases, pests, and weeds, and the number of pesticide applications in the demonstration villages were lower than those in the non-demonstration villages, but the yield per 667 m² was 10% lower. This may be caused by other factors besides diseases, pests, and weeds, such as climatic factors, insufficient application equipment, delays in the optimal application time, and uneven distribution of water and fertilizer due to different terrain heights of different plots during uniform fertilization. In 2018 and 2019, the average number of times of drug use in wheat demonstration villages and non-demonstration villages were three and four times, respectively, and the average number of times of drug use in non-demonstration villages was one more than that in demonstration villages. The average cost of medication per 667 m² in the demonstration villages was ¥24, and that in non-demonstration villages was ¥36.5 per 667 m². The cost of medication per 667 m² in the demonstration villages was ¥12.5 lower than that in the non-demonstration villages.

(4) The assessment report of crop yield impact

In 2015, the average rice yield per 667 m² in the demonstration villages was 599 kg, and the average yield per 667 m² in non-demonstration villages was 425 kg. The average yield per 667 m² in the demonstration villages was 174 kg higher than that in the non-demonstration villages. It should be noted that there was a large farmer with a yield of 600 kg per 667 m² in the demonstration village, and the yield per 667 m² of other growers was 400–450 kg, which was the same as the average yield of rice per 667 m² of growers in non-demonstration villages. In 2016, the average rice yield per 667 m² in the demonstration villages was 525 kg, the average yield per 667m² in non-demonstration villages was 629 kg, and the average yield per 667m² in the demonstration villages was 104 kg lower than that in the non-demonstration villages. In 2017, the average rice yield per 667 m² in the demonstration villages was 600 kg, and the average yield per 667 m² in non-demonstration villages was 540 kg. The average yield per 667 m² in demonstration villages was 60 kg higher than that in non-demonstration villages. In 2018, the average rice yield per 667 m² in the demonstration villages was 620 kg, and the average yield per 667 m² in non-demonstration villages was 530 kg. The average yield per 667 m² in the demonstration villages was 90 kg higher than that in the non-demonstration villages.

In 2019, the average rice yield per 667 m² in the demonstration villages was 598 kg, and the average yield per 667 m² in non-demonstration villages was 521 kg. The average yield per 667 m² in demonstration villages was 77 kg higher than that in the non-demonstration villages. It should be noted that there was only one large planting household in the demonstration village, with a yield of 598 kg 667 m².

In 2015, the average yield of wheat in the demonstration villages was 378 kg per 667 m², the average yield of non-demonstration villages was 478 kg per 667 m², and the average yield of demonstration villages was 99 kg per 667 m² lower than that in non-demonstration villages. In 2016, the average yield of wheat in the demonstration village was 410 kg per 667 m², and the average yield in the non-demonstration village was 355 kg per 667 m². Compared with the two yields, the average yield of the demonstration village was 55 kg per 667 m² higher than that in the non-demonstration village. In 2017, the average yield of wheat in the demonstration villages was 300 kg per 667 m², the average yield of non-demonstration villages was 325 kg per 667 m², and the average yield of demonstration villages was 25 kg per 667 m² lower than that in the non-demonstration villages. In 2018, the average yield of wheat per 667 m² in demonstration villages was 330 kg, and the average yield per 667 m² in non-demonstration villages was 310 kg. The average yield per 667 m² in demonstration villages was 20 kg higher than that in the non-demonstration villages. In 2019, the average yield per 667 m² of wheat in the demonstration villages was 345 kg, and the average yield per 667 m² in the non-demonstration villages was 308 kg. The average yield per 667 m² in demonstration villages was 37 kg higher than that in the non-demonstration villages.

According to investigation and analysis, the project area had produced different degrees of impact on crop yields through the integration and demonstration of key technologies for crop production, emission reduction, and carbon increase, the innovation and application of supporting policies, and the expansion and improvement of public knowledge. Compared with the non-project area, the yield of the project area was higher than that in the non-project area.

The average yield of rice in the demonstration villages in 2015 was higher than that in the non-demonstration villages, indicating that the demonstration technology had been better promoted. In 2016, the yield of rice per 667 m² in non-demonstration villages was higher than that in demonstration villages, because non-demonstration villages had fewer pests and diseases than demonstration villages. The average yield of rice in the demonstration villages from 2017 to 2019 was higher than that in non-demonstration villages, indicating that the demonstration technology had been well promoted in the project area.

In 2015, the average wheat yield in the demonstration villages was lower than that in non-demonstration villages. Based on the occurrence of diseases and insect pests, this was due to the higher incidence of diseases and insect pests in the non-demonstration villages than in the demonstration villages. In 2016, the yield of wheat in the demonstration villages was higher than that in the non-demonstration villages. The average wheat yield in the demonstration villages in 2017 was lower than that in the non-demonstration villages. This may be caused by other factors besides diseases, pests, and weeds, such as climatic factors, insufficient pesticide application equipment in the demonstration area, delays in the optimal application time, and uneven distribution of water and fertilizer due to the different terrain heights of different plots in the demonstration area. In 2018 and 2019, the average yield of wheat in the demonstration villages was higher than that in the non-demonstration villages, indicating that the demonstration technology had been better promoted in the project area.

The project had affected crop yields to varying degrees, mainly affecting the average yield and yield per 667 m^2 of different crops in the demonstration villages and non-demonstration villages, and there was no stable crop yield relationship between the demonstration villages and non-demonstration villages. The two yield theories of individual farmers in the project area are contrary to the actual situation. The possible reasons for this situation are (i) natural factors, such as floods, droughts, and sudden weather effects; (ii) human factors, such as farmers' low enthusiasm for accepting new technologies, and continuing to plant according to the original planting technology; lack of comprehensive quality of farmers, and incomplete understanding of national policies or new technologies. These prevent the technology from being promoted as expected. In view of these possible factors, we should proceed from both objective and subjective aspects, and further promote new technologies to increase crop yields and achieve sustainable agricultural development.

(5) Investigation report on the use of medical devices
(i) The usage of rice medicine equipment. In 2015, both rice demonstration villages and non-demonstration villages used conventional sprayers to spray pesticides. Under the guidance of project experts, farmers in the demonstration villages increased the number of pesticides used, and the yield per 667 m^2 increased by 40% compared with non-demonstration

villages. The project implementation effect was significant. In 2016, both the rice demonstration villages and non-demonstration villages used conventional sprayers. The average yield per 667 m^2 in 2016 of the demonstration villages was 13% lower than that in 2015, and the yield per 667 m^2 in 2016 of the non-demonstration villages was 48% higher than that in 2015. It was estimated that factors (droughts, floods, and other meteorological disasters) other than pesticides in 2016 would have a greater impact on rice production in demonstration and non-demonstration villages. In 2017, 2018, and 2019, the pesticide application method used in the rice demonstration area of the rice demonstration village was a large-scale wide-width sprayer, while the non-demonstration area used an ordinary knapsack electric sprayer. The results measured in the experiment in Huaiyuan County, Anhui Province, showed that the effective use rate of pesticides (56.6%) using large sprayers was higher than that of ordinary knapsack electric sprayers (52.4%). It should be noted that in 2017 and 2018, there was only one large planting household in the demonstration village, with a planting area of 600 ha. The data in the demonstration area was based on the data of this large planting household.

(ii) The usage of wheat medicine equipment. In 2015, both the wheat demonstration villages and non-demonstration villages used conventional sprayers for pesticide application. The frequency of occurrence of diseases, pests, and weeds, and the use of pesticides in the demonstration villages were lower than those of the non-demonstration villages, but the yield per 667 m^2 was 20% lower. Besides diseases, pests, and weeds, other factors had a greater impact on yield. In 2016, both the wheat demonstration and non-demonstration villages used conventional sprayers for pesticide application. The frequency of wheat diseases, pests, and weeds in the demonstration villages was one time lower than that in the non-demonstration villages, the frequency of application were 1.5 times higher, and the yield per 667 m^2 was 16% higher. The project implementation effect was obvious. In 2017, 2018, and 2019, the application method used for wheat in the demonstration area of Huaiyuan County, Anhui Province, was a large-scale wide-width sprayer, while the wheat farmers in the non-demonstration area used ordinary knapsack electric sprayers to spray. The test results in Huaiyuan County, Anhui Province, showed that the effective use rate of pesticides (56.6%) using large-scale wide-width pesticide applicators was higher than that of ordinary knapsack electric sprayers (52.4%) in non-demonstration areas. The results showed that to achieve the same control effect, the average number of pesticides in non-demonstration areas should be one more time than in demonstration areas, which makes the average pesticides cost per 667 m^2 (¥34.5) in non-demonstration areas also higher than the cost of pesticides in demonstration villages (¥23.6/667m^2).

(6) Questions and suggestions

  (i) The impact of non-project factors on project implementation effects and suggestions. The impact of project crop yields may be affected by farmers' planting habits, floods, droughts, various meteorological disasters, and human factors. The evaluation of project implementation effects needs to consider the comprehensive impact of various factors. It is suggested that other factors affecting crop yield should be taken into consideration, and the impact of non-project factors on the implementation effect of the project should be stripped off so as to test the implementation effect of the project.

 (ii) Suggestions on project implementation: During the investigation, it is found that drugs used by farmers in demonstration villages and non-demonstration villages are more complicated. It is necessary to test the rationality of the drug used by farmers in demonstration villages and non-demonstration villages based on actual conditions in future project implementation.

(iii) Selection of research objects: Omitting the number of households investigated during the implementation of the project, the neatness of the demonstration villages and non-demonstration villages are poor during the year, the number of households varies, and the scale of planting varies, so it is difficult to make a scientific comparison. It is recommended to select fixed farmers for long-term monitoring to improve the scientific nature of project research.

(iv) The usage of medical equipment: The newly purchased large-scale wide-width pesticide applicators and drones in the project area should be used on a large scale as soon as possible to conduct investigations on the efficacy and use of new devices in the project area.

7.3.4 *Monitoring and Evaluation of Social Impact*

(1) Project background and research situation

As the international community pays more and more attention to climate change, greenhouse gas emission reduction, and food security, research on farmland soil carbon sequestration and emission reduction technology has received unprecedented attention from the scientific community and has gradually been widely used in agricultural production in various countries. The climate-smart agriculture aims to improve the nitrogen fixation capacity of farmland soil without reducing the level of crop productivity while reducing field greenhouse gas emissions and adopt diverse cultivation management modes to enhance crop production's adaptability to climate change.

China's climatic conditions, land resources, and planting systems all have obvious regional characteristics. Carbon sequestration and emission reduction technologies have different requirements and effects in various regions, and certain management measures are difficult to be promoted continuously due to their impact on production. Wheat, rice, and corn are the three-grain crops in my country, and their total yield accounts for more than 85% of China's grain yield. The main grain production areas in North China and East China are responsible for ensuring food security. The sown area and grain yield of grain crops account for 63% and 67% of the country's total grain crop area and yield, respectively. At the same time, major food production areas are also facing serious organic carbon losses, large amounts of nitrogen fertilizers, and other problems, as well as urgent needs for carbon sequestration and huge potential for energy saving and emission reduction of greenhouse gases. Thus, the promotion and application of energy-saving and carbon sequestration technologies under the premise of ensuring food production in major grain production areas, as well as demonstration and evaluation of emission reduction effects, can not only improve soil fertility and productivity, and slow down the emissions of greenhouse gases in the soil, but also maintain agriculture in my country. It is also a strategic choice for sustainable development.

Huaiyuan County is located in northern Anhui, the middle reaches of the Huai River (32°43'–30°9'N, 116°45'–117°19'E). The county has a total area of 2,212 square kilometers and a total population of 1.26 million. The project area of Wanfu Town is located in the southwest of Huaiyuan County, about 40 kilometers away from the county seat. The project area is entirely a plain area with two rivers in it, namely the Qian River and the Cihuaixin River. The annual average temperature is 15.8°C, the highest monthly (July) average temperature is 28.6°C, the lowest monthly (January) temperature is 1.5°C; the annual precipitation is 791.6 mm, the highest monthly precipitation (July) is 188.1 mm, the lowest monthly precipitation (January) is 7.6 mm; the accumulated sunshine hours are 1,889.8 hours, the largest monthly accumulated sunshine hours

(April) is 207.7 hours, and the lowest monthly accumulated sunshine hours (February) is 90.7 hours. The frost-free period averages 219 days.

In 2017, we visited two project villages in Huaiyuan, Anhui Province, namely Zhuanqiao Village and Liulou Village. Also, two large outdoor households, to whom 40 households transferred their land were selected for investigation. To conduct a comparative analysis of project effects, five non-project villages were selected, namely Sunzhuang Village, Zhaomu Village, Chen'an Village, Zhendong Village, and Zhennan Village. Each village selected 15 farmers for the investigation. The project team members communicated with village officials and filled out village-level questionnaires, and learned in detail about the socioeconomic conditions, crop planting conditions, land circulation conditions, and cooperative development conditions of the seven villages, and received a total of 117 valid questionnaires. In 2018, the survey focused on eight villages in Wanfu Town and Lanqiao Town in Huaiyuan County, including four project villages in Wanfu Town, namely Liulou Village, Zhuanqiao Village, Zhennan Village, and Zhendong Village; the other four non-project villages, namely Huimin New Village, Sunzhuang Village, Zhaomu Village, and Chen'an Village, received a total of 97 available questionnaires. In 2019, the project team members investigated the impact of climate-smart major food crop production projects on local socio-economic development and ecological development in Huaiyuan County, Anhui Province. The survey focused on eight villages in Wanfu Town, Huaiyuan County, including Zhennan Village, Zhendong Village, and Zhaomu Village. A total of 100 available questionnaires were collected. Eight project team members went to Huaiyuan County, Anhui Province, to investigate the impact of climate-smart major food crop production projects on local socioeconomic development and ecological development.

(2) Evaluation of the economic and social effects of project implementation
(i) Stabilize production and reduce input, and promote increased production and income

The yield of major food crops rose steadily. From 2016 to 2019, the average rice yield in the project area and non-project area of Huaiyuan County was 8,009 and 8,004 kg ha^{-1}, respectively. The rice yield in non-project areas showed a downward trend, and the decline rate was about 230 kg ha^{-1} yr^{-1}. Since the implementation of the project in 2016, the rice yields in the project areas from 2017 to 2019 were higher than those in the non-project area. From 2016 to 2019, the wheat yield in the project area and the non-project area of Huaiyuan County decreased first and then recovered. The average wheat yield in the project area and the non-project area was 4,685 kg ha^{-1} and 4,413 kg ha^{-1}, respectively. The average yield of wheat in the project area was higher than that in the non-project area. From 2016 to 2018, the difference in wheat yield between the project area and the non-project area increased at an average annual rate of 250 kg ha^{-1}, and there was a slight decrease in 2019. From 2016 to 2019, the yield per unit area in the rice–wheat rotation system project area and non-project area of Huaiyuan County increased by 374 kg ha^{-1} and 76 kg ha^{-1} yr^{-1} respectively, and the 4-year average yield was 12,695 and 12,418 kg ha^{-1}, respectively. Since the implementation of the project in 2016, the annual wheat–rice yields in the project areas from 2017 to 2019 were higher than those in the non-project areas, with an average of about 615 kg ha^{-1} higher.

The planting costs of major food crops have gradually decreased. From 2016 to 2019, the cost of rice planting in the project area and non-project area of Huaiyuan County decreased by an average of ¥52 and ¥79 ha^{-1} respectively, and the 4-year average cost was ¥8,350 and ¥8,390 ha^{-1}, respectively. The planting cost in the project area was relatively low. From 2016 to 2019, the cost of wheat planting in the project area and non-project area of Huaiyuan County decreased by ¥107 and ¥112 ha^{-1} respectively, and the 4-year average cost was ¥6,500 and ¥6,618 ha^{-1}, respectively. The project area was ¥118 ha^{-1} lower than the non-project area on average. From 2016 to 2019, the annual wheat–rice planting cost in the project area and non-project area of Huaiyuan County was decreased by an average of ¥159 and ¥191 ha^{-1}, respectively, and the 4-year average cost was ¥14,850 and ¥15,010 ha^{-1}, respectively. The project area was ¥160 ha^{-1} lower than the non-project area on average.

The decline in the production cost of major food crops reflects the positive results of the climate-smart food project; however, agricultural production costs are still at a relatively high level. The higher agricultural production input in the project area, on the one hand, guarantees the output of grain, and, on the other hand, it also increases the expenditure burden of farmers and restricts the continuous increase of agricultural operating income.

The income per unit area of major food crops is stable and improving. From 2016 to 2018, the income per unit area of rice and wheat in Huaiyuan County showed a downward trend, and both rebounded significantly in 2019. The 4-year average rice yield per unit area in the project area and the non-project area was ¥11,706 and ¥11,603 ha^{-1}, respectively; since the implementation of the project in 2016, the rice yield per unit area in the project area from 2017 to 2019 was higher than that in the non-project area, with a 3-year average ¥780 ha^{-1}. From 2016 to 2019, the average yield per unit area of wheat in the project area and non-project area was ¥2,835 and ¥2,226 ha^{-1}, respectively, and the yield per unit area in the project area was higher than that in the non-project area, with an average increase of about ¥609 ha^{-1} in 4 years. From 2016 to 2018, the annual wheat–rice crop rotation system in project area and the non-project area changed little, but in 2019, it increased by 57.4% and 68.0%, respectively, compared with 2018. The 4-year average income per unit area of the project area and the non-project area was ¥14,540 and ¥13,829 ha^{-1}, respectively. The income per unit area of the project area was about ¥710 ha^{-1} higher than that in the non-project area.

The degree of agricultural mechanization has gradually increased, and the production efficiency has been greatly improved. In recent years, with the support of the project and national agricultural machinery subsidies, the mechanization rate of project counties has continued to increase. The total power of agricultural machinery in Huaiyuan County has reached 2.77 million kilowatts. There are 175,000 tractors of various types, including about 5,500 large and medium-sized tractors and 6,400 combine harvesters. The structure of agricultural machinery is continuously optimized and the level of equipment continues to improve.

With the continuous increase of agricultural machinery, agricultural socialized services in the form of agricultural machinery cooperatives and individuals have flourished. According to the survey situation, the machine farming, machine seeding, and machine yield rate of rice in Huaiyuan County are above 90%, and the machine farming, machine seeding, and machine yield rate of wheat are above 85%. The Difa Agricultural Technology Company even used advanced agricultural production equipment such as drones for production, which greatly improves efficiency and saves a lot of labor. The mechanization of agricultural production and social services improved the efficiency of agricultural production, while greatly reducing the labor burden of farmers, saving labor time, and increasing farmers' social welfare.

New planting and breeding models are being developed during exploration. With the exploration and implementation of new farming models such as shrimp farming in rice fields and duck farming in rice fields, the increase in agricultural production and efficiency has been further promoted. Raising ducks in paddy fields is only supplemented in the early stages of duck growth, and in the later stages, ducks are fattened by weeds and pests in water. The crops do not need to be weeded or sprayed with pesticides, which is environmentally friendly. Duck grazing in the wild not only has low cost and good meat quality but also has a broad market prospect. In 2018, the Difa Agricultural Technology Company of Anhui Province carried out duck breeding in 200 ha of rice fields and sold 3.3 million catties of new varieties of Nanjing rice at a price of ¥0.9–1.4 per kg, sales reached ¥7.2 million, and food income was much higher than ordinary rice cultivation. Raising 30,000 ducks in the paddy field required only 130 days of growth. The sales price of ducks was ¥4 per kg, the net profit of ducks was ¥12 per duck, and the net income of ducks was ¥360,000. Also, the company had also launched a new type of rice and shrimp breeding model, raising prawns on 33.33 ha of rice and wheat, and each 667 m^2 of rice could increase the sales of prawns by ¥5,000, and the net profit of prawns per acre would reach ¥2,900. With the support and promotion of the project, the project area is constantly exploring and improving new models such as rice loach and lotus root duck. Not only has it achieved a double harvest of economic and ecological benefits, but it has also achieved a multi-win situation for farmers, agricultural enterprises, and village collectives. It provides better ideas and valuable experience for the development of the combined model of rural planting and breeding and the increase in agricultural production and efficiency.

(ii) Reducing the usage of fertilizers and pesticides, promoting energy saving and emission reduction

The use of chemical fertilizers for major food crops was decreasing year on year. From 2016 to 2019, the fertilizer consumption of rice in Huaiyuan County increased first and then decreased. The average fertilizer consumption for rice in the project and non-project areas in 2016–2019 was 1,065 and 1,059 kg ha^{-1}, respectively. From 2016 to 2019, the difference in the use of rice fertilizer between the project area and the non-project area had decreased year on year, at a rate of 22 kg ha^{-1} yr^{-1}. Starting in 2017, the use of rice fertilizer in the project area would be equal to that in the non-project area. After 2018, the use of rice fertilizer in the project area was less than that in the non-project area. From 2016 to 2019, the use of wheat fertilizer in Huaiyuan County showed a downward trend. The project area and non-project area decreased at a rate of 26 and 4 kg ha^{-1} yr^{-1}, respectively. The average use of wheat fertilizer in the project area and non-project area from 2016 to 2019 was, respectively, 980 and 960 kg ha^{-1}, and the difference in the use of chemical fertilizers for wheat in the project area and non-project areas from 2016 to 2019 have been reduced year on year, at a rate of 20 kg ha^{-1} yr^{-1}. Starting in 2019, the use of chemical fertilizer for wheat in the project area was less than that in the non-project area. In the rice–wheat rotation system of Huaiyuan County from 2016 to 2019, the fertilizer consumption showed a downward trend. The 4-year average fertilizer consumption in the project and the non-project areas was 2,047 and 2,016 kg ha^{-1}, respectively. Although fertilizer consumption in the project area was high, the gap with the non-project area was decreasing year on year, and there was a steady and positive development trend.

On the whole, the reduction in the use of chemical fertilizer reflects the successful demonstration and application of the key part of the project — the reduction of agricultural fertilizer application technology. However, for the absolute quantity, the use of chemical fertilizer in the project area is still relatively large, and further reductions are needed.

The use time of pesticides in major food crops was decreasing year on year. From 2016 to 2019, the number of pesticide applications for rice in Huaiyuan County showed a significant downward trend. Project and non-project areas decreased at a rate of 0.3 and 0.2 times per year, respectively, and the average number of applications for 4 years was 5.3 times. During the 4 years, the number of pesticide use between the project area and the non-project area had gradually decreased. By 2017, the average number of pesticide use in the project area was lower than that in the non-project area. From 2016 to 2019, the number of wheat pesticide applications in Huaiyuan County showed an overall downward trend. The project area experienced a significant decrease of 0.1 times per year, while the non-project area declined significantly from 2016 to 2018, with an average annual decrease of 0.2 times over the 3 years. The average decline in the 4 years from 2016 to 2019 was small, only 0.1 times per year. From 2016 to 2019, the number of wheat medications between the project area and the non-project area decreased at a rate of 0.1 times per year, and the decline was rapid from 2017 to 2019. From 2016 to 2019, the number of medications used in the wheat–rice rotation system in Huaiyuan County showed a significant decrease. Project areas and non-project areas decreased at a rate of 0.5 and 0.3 times per year, respectively. The average number of medications in the 4-year period was 9.3 and 9.4, respectively.

In the past 4 years, the gap in the number of medications between the project area and the non-project area had gradually decreased. In 2018, the number of medications in the project area was slightly lower than that in the non-project area.

After implementation of the climate-smart major food crops project in Huaiyuan County during 2016 and 2019, the amount of fertilizer used in the rice–wheat rotation system had basically remained stable. The use of non-project fertilizers had continued to increase at a rate of 64 kg ha^{-1} yr^{-1}, and the number of pesticides used in the project area of the system had decreased at a faster rate than in the non-project area.

The returning rate of straw to the field has increased, and farmers have increased their willingness for green and sustainable development. Returning straw to the field is a key technology of the climate-smart main food crop project. It is not only a measure to increase the production of fertility but also a core technology for carbon sequestration and emission reduction, with obvious emission reduction effects. The straw contains a large amount of fresh organic materials, which can be converted into organic matter and quick-acting nutrients after a period of decomposition after being returned to the farmland. Returning straw to the field can increase soil organic matter, improve soil structure, make the soil loose, increase porosity, reduce capacity, promote microbial activity and the development of crop roots, improve soil's physical and chemical properties, and can also supply potassium and other nutrients. Returning straw to the field can promote agricultural water-saving, cost-saving, increase in production, and efficiency. It also has significant effects on increasing fertilizer and yield, generally increasing the yield by 5–10%. Since the implementation of the project, the project area has achieved a 90% returning rate, which is very effective.

Farmers obtained a better understanding of science, technology, and environmental protection, and their willingness to save energy and reduce emissions has also continued to increase. The survey in Huaiyuan County showed that 81% of the farmers indicated that they would accept a unified fertilization formula; 80% of the farmers were willing to accept a unified pest control; 75% of the farmers indicated that climate change had an impact on agricultural production, and most of them were negative; 83% of the farmers believed that agricultural technology was very important for solving current agricultural problems. On the whole, farmers in the project area obtained increasingly better awareness of agricultural science, technology, and environmental protection. They have accepted the concept of climate-smart food crop projects from a cognitive perspective and are willing to follow them up through action.

Drive disadvantaged groups to help fight poverty. Climate-smart food projects actively integrate the national rural revitalization strategy and targeted poverty alleviation work and attach great importance to the care of vulnerable groups. During the implementation of the project, researchers provided the greatest help and encouragement to vulnerable groups including women, the elderly, the disabled, and poor households.

In Huaiyuan County, farmers who participated in the climate-smart main food crops project accounted for 7.5% of the poor households; among those farmers, the poor households accounted for 18%; among the trained personnel, the poor accounted for about 23%, and the proportion of women was as high as 58%. The average number of employed persons with disabilities in the village was 14, and their average income reached more than ¥3,820; the average number of women employed in the village was 250, and the average income reached ¥4,300, the average number of employed elders in the village was 330, and the average income reached more than ¥4,000.

Chapter 8

Practice and Exploration of Wheat–Maize Production with Climate-Smart Agriculture in China

Abstract

The practice of climate-smart wheat–maize production technology mainly focuses on new materials for carbon sequestration and emission reduction, new patterns for carbon sequestration, and cultivation techniques for conservation tillage. The positive effects of urease inhibitors, nitrification inhibitors, biochar fertilizers, and other new carbon sequestration materials in reducing greenhouse gas emissions and improving resource utilization efficiency were illustrated through experiments and model integration demonstration in the past 5 years. The effects of conservation tillage on carbon sequestration, emission reduction, and yield stabilization and efficiency enhancement in the wheat–maize system were clarified. The effects of carbon sequestration, emission reduction, and steady yield increase in new climate-smart agricultural production modes such as winter wheat-summer peanut and winter wheat-summer soybean were explored. The popularization and application of climate-smart agriculture ideas and technologies are promoted through activities such as training of climate-smart agriculture ideas and related technologies, training related to crop high-yield and high-efficiency management techniques and guidance, crop production conservation tillage technology, and crop green-integrated control technology. By carbon sequestration and emission reduction, environmental effect, pest management, and social impact of monitoring and evaluation, through 5 years of follow-up investigation, the differences of pest and disease frequency, pesticide use, mechanization degree, crop yield, agricultural production cost and soil carbon storage between farmers of wheat–maize planting system in the project area and the non-project area were systematically analyzed, the application effect of climate-smart agriculture was explained, and the feasibility and social influence of climate-smart agriculture were quantitatively evaluated. In general, the implementation of climate-smart projects has strongly promoted carbon sequestration and emission reduction of crop production, environmental green ecology, and stable production and income increase of farmers in the project areas.

8.1 Practice of Climate-Smart Wheat–Maize Production Technology

In order to improve the land quality of the wheat–maize production system and reduce greenhouse gas emissions in agricultural production, the project team carried out screening and demonstration tests of new materials for carbon sequestration and emission reduction, new model tests for carbon sequestration and emission reduction, and tests of conservation tillage and supporting cultivation techniques in Ye County from 2015 to 2018. Among them, the new materials mainly include biochar fertilizer (BF), nitrification inhibitor dicyandiamide (DCD), urease inhibitor hydroquinone (HQ), and polymer-coated urea (PCU). The new model mainly includes wheat–maize, wheat–peanut, and wheat–soybean rotation. Conservation tillage and supporting cultivation techniques mainly include no-tillage (NT), rotary tillage (RT), and conventional tillage (CT), as well as no nitrogen (NN), low nitrogen (LN), and conventional nitrogen (CN) fertilization.

Soil enzyme activity, nitrogen mineralization rate, greenhouse gas emission, and crop yield were analyzed. The results showed that: (1) the application of urease inhibitor (HQ) and nitrification inhibitor (DCD) could reduce the activity of soil sucrase and urease, while the application of polymer-coated urea (PCU) and biochar fertilizer (BF) could improve the activity of soil sucrase; (2) urease inhibitors (HQ) and application of nitrification inhibitor (DCD) can reduce the amount of ammonium nitrogen in the soil to the transformation of nitrate nitrogen, with urease inhibitors (HQ) and nitrification inhibitor (DCD) providing the best inhibition effect; compared with conventional fertilization (U), the treatment of soil NH_4^+-N concentration increased by 65.18%, while NO_3^+-N concentrations were reduced by 56.14%; (3) the use of various new materials for carbon sequestration and emission reduction will reduce the greenhouse gas emission of winter wheat to a certain extent, and the reduction effect of polymer-coated urea (PCU) is the most obvious; (4) the use of nitrification inhibitors (DCD) and biochar fertilizer (BF) will increase wheat yield; (5) wheat–peanut and wheat–soybean rotation modes can effectively reduce greenhouse gas emissions in farmland compared with wheat–maize rotation mode; and (6) conventional tillage (CT) is beneficial to improving the activity of soil sucrase and urease. The soil sucrase activity could be improved by increasing nitrogen application, but it had no significant effect on urease activity; (7) the no-tillage+conventional nitrogen application (NT+CN) and conventional tillage+low nitrogen (CT+LN) had the best emission reduction effect; (8) under rotary tillage (RT) and no-tillage (NT), the net nitrogen mineralization rate of soil was high. The net nitrogen mineralization rate of soil can be decreased by reducing nitrogen fertilizer use; (9) the conventional tillage+conventional nitrogen application (CT+CN) is not only beneficial to the emergence of wheat seedlings but also to the higher yield of wheat.

To sum up, it is of great practical significance for the development of climate-smart agriculture (CSA) in this region to use appropriate new materials for carbon sequestration and emission reduction, appropriately increase the planting area of peanut and soybean, and combine reasonable tillage measures with cultivation regulation measures.

8.1.1 *Screening and Demonstration of New Materials for Carbon Sequestration and Emission Reduction in Farmland*

The application amount of nitrogen fertilizer is increasing year on year, which is an important problem in the actual agricultural production in China. After urea is applied to the soil, under the action of urea hydrolytic specific enzyme, urease, a large amount of ammonium nitrogen can be converted into ammonium in a short period of time, leading to serious ammonia volatilization and fertilizer loss, and finally resulting in the increase of greenhouse gas emissions in farmland. At present, the research on nitrogen use efficiency of control measures mainly focused on adding urea enzyme/nitrification inhibitors to improve stability. Research on the coated fertilizer-controlled release urea method and its influence on the previous studies are often limited to a single type of fertilizer research, and only few of the different nitrogen fertilizer regulation measures were analyzed. Therefore, the study and analysis of different nitrogen fertilizer regulation measures and the screening of new ways that can reduce green-house gas emissions and reduce the negative effect of nitrogen fertilizer application are of great significance for the construction of farmland eco-logical civilization and the mitigation of global warming trend. In this study, the effects of urease/nitrification inhibitor addition and envelop-ment fertilizer application on soil nitrogen conversion, greenhouse gas emission, and wheat yield in wheat fields were studied by optimizing nitrogen fertilizer management measures in the field, so as to provide a theoretical basis for the development of fertilizer models for greenhouse gas emission reduction in farmland in this region.

The project team began to carry out screening tests of new materials for carbon sequestration and emission reduction in Quanji Village, Longquan Township, Ye County, Henan Province, in October 2015, and will demon-strate the application of new materials from October 2017 to October 2019 (Figure 8.1). In the project area, representative plots were selected to set

Figure 8.1.   Screening and demonstration of new materials for carbon sequestration and emission reduction in farmland in wheat–maize production system.

corresponding field positioning experiments to compare the effects of straw returning and biochar and other new materials on crop growth and development, yield, greenhouse gas emissions, and soil carbon and nitrogen changes under the wheat–maize planting mode. New materials in this study include biochar fertilizer (BF), nitrification inhibitor dicyandiamide (DCD), urease inhibitor hydroquinone (HQ), and polymer-coated urea (PCU). The dosage of biochar unit is 0.5 kg m^{-2} (5,000 kg/ha, straw biochar into quantity throughout the year), the nitrification inhibitor dicyandiamide unit dosage is 1 g m^{-2} (10 kg/ha), the dosage of urease inhibitor hydroquinone unit 0.1 g m^{-2} (1 kg/ha), and the dosage of polymer-coated urea unit 45 g m^{-2} (450 kg/ha). Four kinds of nitrogen fertilizer control measures were set, namely urea+nitrification inhibitor (U+DCD), urea+urease inhibitor (U+HQ), and urea+urease inhibitor+nitrification inhibitor (U+HQ+DCD), and routine routine fertilization (U) was used as the control. The other production management measures were consistent with local routine.

In 2018, six kinds of nitrogen fertilizer control measures were set based on straw returning to the field by using the method of field plot test, including: routine fertilization (U), urea+nitrification inhibitor (U+DCD), urea+urease inhibitor (U+HQ), urea+nitrification inhibitor+urease inhibitor (U+DCD+HQ), polymer-coated urea (PCU), and biochar fertilizer (BF). The tested crop was winter wheat (Zhoumai 27). Taking equal nitrogen (pure nitrogen 225 kg ha^{-1}) as the fertilization principle, the application amounts of nitrogen (N), phosphorus (P_2O_5), and potassium (K_2O) in each treatment were respectively 225, 75, and 150 kg ha^{-1}. The inhibitors were applied uniformly mixed with fertilizer, and the application amounts of urease inhibitor (HQ) and nitrification inhibitor (DCD) were, respectively, 0.5% and 5% of the urea dosage. Nitrogen fertilizer was applied two times, 60% as basal fertilizer (October 12, 2018), and the remaining 40% as topdressing during wheat jointing stage (March 31, 2018). In December 2017, and in February, March, April, and May 2018, the investigation of soil characters and the determination of relevant indicators at various growth stages of wheat were completed.

(1) Effects of different nitrogen fertilizer regulation measures on soil enzyme activity

The activity of sucrase is highly correlated with soil organic matter, nitrogen, and phosphorus content, microbial quantity, and soil respiration. The experimental results in 2016 and 2017 showed that the application of urease inhibitor (HQ) and nitrification inhibitor (DCD) could reduce the activity of soil sucrase, while the application of polymer-coated urea (PCU) and biochar fertilizer (BF) could improve the activity of soil sucrase. In different growth stages, the activity of soil sucrase treated with urea+urease inhibitor (U+HQ) was the lowest among all treatments, indicating that urease inhibitor (HQ) played a good role in inhibiting sucrase. Urea+nitrification inhibitor (U+DCD) and urea+nitrification inhibitor+urease inhibitor (U+DCD+HQ) also reduced soil sucrase activity to different degrees after topdressing. After wheat entered the turning green stage, the activity of soil sucrase showed a trend of first increasing and then decreasing under different nitrogen fertilizer control measures, and the peak appeared after topdressing. After topdressing, the concentration of the reaction substrate increased, and the concentration of NH^+_4-N in the soil also increased sharply, which accelerated the decomposition of soil organisms. The content of organic matter in the soil kept increasing, and the activity of soil sucrase also showed an increasing trend.

Urease's enzymatic reaction product, nitrogen, is one of plant nitrogen sources. Urea applied to soil can only be hydrolyzed with the participation of urease. Its activity can reflect the level of soil nutrient supply and reflect the carbon conversion rate and respiration intensity in soil. The results showed that there were differences in soil urease activity under the action of different new materials, but they all reduced the soil urease activity to different degrees, among which the addition of urease inhibitor (HQ) had the greatest effect on the reduction of soil urease activity. From the turning green period to the jointing period, the activity of soil urease decreased in all treatments, indicating that wheat grew rapidly after entering the turning green period and continuously consumed soil nutrients, which led to the decrease of urease activity. With the supply of nutrients after topdressing, wheat grew vigorously, underground roots developed, and biological and microbial activities flourished, which stimulated the activity of urease. Soil urease activity in each treatment increased rapidly and then tended to be stable and began to decline after 4–20 days.

(2) Effects of different nitrogen fertilizer regulation measures on soil mineral nitrogen

Among different nitrogenous fertilizer regulation measures, urea+urease inhibitor (U+HQ), urea+nitrification inhibitor (U+DCD), and urea+nitrification inhibitor+urease inhibitor (U+DCD+HQ) can all reduce the soil nitrate concentration and play a good role in nitrogen regulation. Test results showed that compared with conventional fertilization (U), adding inhibitor treatment of soil NO^-_3-N content at the same time was low, which showed that urease inhibitors (HQ) and nitrification inhibitor (DCD) can inhibit the soil ammonium nitrogen to the transformation of nitrate nitrogen oxide and decrease soil NO^-_3-N concentrations. The inhibition effect of urea inhibitor (HQ) and nitrification inhibitor (DCD) was the best, and the maximum decrease rate of NO^-_3-N concentration was 56.14%. In the process of wheat growth, the concentration of NO^-_3-N in all treatments increased from fertilization and sowing to the peak, and gradually declined with the advance of wheat growth period. Finally, the concentration of NO^-_3-N reached almost the same level in the mature stage. After urea is applied to the soil, it is decomposed to NH^+_4-N under the action of urease, and NH^+_4-N is easily oxidized to NO^-_3-N. At this time, the content of nitrate nitrogen in the soil is greatly increased. However, after the accumulation of NO^-_3-N in soil, soil NO^-_3-N under different nitrogen fertilizer measures was reduced to a similar level through various channels such as leaching, absorption, and denitrification loss.

The test results in 2017 showed that urea+urease inhibitor (U+HQ), urea+nitrification inhibitor (U+DCD), and urea+nitrification inhibitor+ urease inhibitor (U+DCD+HQ) could increase the concentration of ammonium nitrogen in soil, reduce the concentration of nitrate in soil, and play a good role in nitrogen regulation. Compared with conventional fertilization (U), the soil treated with urea+urease inhibitors (U+HQ) and urea+nitrification inhibitor (U+DCD) had higher NH^+_4-N content in winter wheat seedling stage. The results also showed that urease inhibitor (HQ) and nitrification inhibitor (DCD) played a good role in improving the concentration level of NH^+_4-N in soil. The urea+nitrification inhibitor+urease inhibitor (U+DCD+HQ) mode also increased the content of NH^+_4-N, so that the concentration of NH^+_4-N in the soil remained at a high level during the maturing stage of wheat. Under different nitrogen fertilizer control measures, soil NH^+_4-N concentration peaks after topdressing. Both the ambient temperature and soil moisture fully meet the decomposition conditions of urea, leading to a significant increase in soil NH^+_4-N content, and nitrogen fertilizer regulation measures help preserve soil NH^+_4-N for a long time. Compared with conventional fertilization (U), soil NH^+_4-N concentration under urea+urease inhibitor (U+HQ), urea+nitrification inhibitor (U+DCD) and urea+nitrification inhibitor+ urease inhibitor (U+DCD+HQ) increased by 49.15%, 86.18% and 65.18%, respectively.

(3) Effects of different nitrogen fertilizer regulation measures on greenhouse gas emission in farmland

(i) N_2O emission: The addition of nitrification inhibitor (DCD) can effectively reduce the emission of N_2O in farmland. The test results showed that the peak value of N_2O emission usually appeared in the base fertilizer period and topdressing period, and the change rate of N_2O emission during jointing top fertilizer period was relatively drastic, which was manifested as emission. Among different new materials for carbon sequestration and emission reduction, the N_2O emission peak value under conventional fertilization mode (U) was the highest (0.18 mg $N \cdot m^{-2} h^{-1}$). The peak value of nitrification inhibitor (DCD) was only 0.11 mg $N \cdot m^{-2} h^{-1}$. After wheat sowing, the soil moisture and temperature conditions were suitable, and the decomposition rate of fertilizer applied into the soil was faster. At this time, the soil nutrient absorption capacity of wheat seedlings was weak, which led to the surplus of nitrogen, thus providing an adequate nitrogen source for the generation of N_2O. With the increase of nitrogen absorption capacity of wheat and the decrease of nitrogen sources, the N_2O emission flux gradually decreased, and the lowest emission flux was only 0.01 mg N $m^{-2} h^{-1}$ in the overwintering period of wheat. After wheat enters the green turning period, with the gradual rise of temperature, the emission of N_2O also increases rapidly after topdressing, and the emission peak appears. Subsequently, the emission of N_2O gradually decreases until the wheat ripens and the emission intensity drops to a low level. In the entire period of wheat growth, urea+nitrification inhibitors (U+DCD), urea+nitrification inhibitors, and urease inhibitors (U+DCD+HQ), compared with the conventional fertilization (U), to a different extent reduced the N_2O emissions, which shows that applying nitrification inhibitor (DCD) will have an inhibitory effect on soil nitrification, thereby reducing N_2O emissions.

(ii) CO_2 emission: The use of various new carbon sequestration and emission reduction materials can reduce the CO_2 emission of soil, among which the reduction effect of polymer-coated urea (PCU) and urease inhibitor (HQ) is significant. CO_2 emission peaks appeared after base fertilizer seeding, and the emission flux peaks appeared during jointing topdressing. Among different new materials for carbon sequestration and emission reduction, the peak value of CO_2 emission under conventional fertilization mode (U) was 619.95 mg $C \cdot m^{-2} h^{-1}$, while the peak value of CO_2 emission under molecular coated urea (PCU) was only 419.41 mg $C \cdot m^{-2} h^{-1}$.

After wheat planting, the CO_2 emission flux was kept at a relatively high level under the influence of fertilizers, land tillage, and other factors. With the decrease of air temperature and soil moisture, the CO_2 emission flux gradually decreased, and the minimum value was only 3.90 mg $C·m^{-2} h^{-1}$ in the overwintering period. After wheat entered the turning green period, the growth rate of wheat plants and roots was accelerated, and meanwhile, the soil microbial activity increased, so did the soil respiration rate, thus promoting the increase of CO_2 emission flux. From jointing stage to flowering stage of wheat after topdressing, CO_2 emission flux peaked, and it presented the order of U (619.95 mg $C·m^{-2} h^{-1}$) > U+DCD (613.83 mg $C·m^{-2} h^{-1}$) > U+HQ+DCD (566.59 mg $C·m^{-2} h^{-1}$) > U+HQ (482.86 mg $C·m^{-2} h^{-1}$) > PCU (419.41 mg $C·m^{-2} h^{-1}$), indicating that polymeric-coated urea (PCU) and urea+urease inhibitor (U+HQ) significantly reduced the peak value of CO_2 emission flux ($p < 0.05$).

(iii) CH_4 emission: Among the new materials for carbon sequestration and emission reduction, the peak value of CH_4 absorption was the highest (0.21 mg $C·m^{-2} h^{-1}$) and the peak value of CH_4 absorption was the lowest (0.10 mg $C·m^{-2} h^{-1}$) under conventional fertilization mode (U). In the entire growth period of wheat, the CH_4 emission curve fluctuated greatly, and the exchange flux showed as absorption. After wheat was just sown, the wheat field had a higher CH_4 absorption flux under different nitrogen fertilizer regulation measures, which may be due to fluffy soil structure after tillage and other measures, sufficient oxygen content in the soil, and the high activity of CH_4 oxidizing bacteria. But the activity of CH_4-producing bacteria was inhibited at this time, and the strong oxidation effect of soil was on CH_4. The CH_4 uptake decreased after irrigation as seen from the CH_4 emission curve/ At higher soil moisture content (19.30%) after irrigation, the effective gap in the soil decreases. Atmospheric CH_4 and O_2 moving into soil is hampered. As the diffusion amount decreases, the lack of reaction substrate and O_2 reduces the activity of CH_4 oxidizing bacteria, which leads to the decrease of CH_4 absorption. At this point, another possible reason for CH_4 absorption decline is that after topdressing, the content of NH^+_4 in soil increases, and too much NH^+_4 competes with CH_4 to compete for the active site of methane oxidase, thus reducing the ability of soil oxidation to absorb CH_4.

(iv) Comprehensive greenhouse gas emissions: Compared with conventional fertilization (U), different nitrogen fertilizer regulation measures effectively reduced the cumulative emission of N_2O, and the emission reduction effect showed the trend of U+DCD>PCU>U+HQ>U+HQ+DCD. The N_2O emissions were reduced by 62.24%, 51.88%, 13.05%, and 5.40%, respectively. Under the conventional fertilization (U) processing, the highest accumulated CO_2 emissions was 5,701.17 kg C ha^{-1}, urea+urease inhibitors (U+HQ), urea+nitrification inhibitors, urease inhibitors (U+DCD+HQ), and PCU significantly reduced CO_2 emissions; urea+nitrification inhibitors (U+DCD) measures can effectively reduce the emission of CO_2 but did not reach a significant level ($p < 0.05$). Under the control measures of polymer-coated urea (PCU), the minimum cumulative CO_2 emission was 3,812.78 kg C ha^{-1}, and the emission reduction efficiency reached 33.12%. The accumulation of CH_4 uptake occurred in U processing fields, showing largest cumulative amount of 4.77 kg C ha^{-1}. While using the other measures for nitrogen fertilizers, CH_4 absorption of farmland was lower than that of U treatment, and they all reached a significant level ($p < 0.05$). The CH_4 absorption of urea+nitrification inhibitor+urease inhibitor (U+DCD+HQ) was the smallest, only 1.53 kg C ha^{-1}. Considering different nitrogen fertilizer regulation measures on soil nitrogen transformation, greenhouse gas emissions and the influence of wheat yield, urea+nitrification inhibitor (U+DCD), and polymer-coated urea (PCU) have obvious advantages over conventional fertilization (U). In the actual production of this area in the future, nitrogen fertilizer combined with nitrification inhibitor and nitrogen fertilizer control measures of coated urea can be used.

Greenhouse gas emissions per unit yield of wheat field by different inhibitors were shown as U+HQ+DCD>U>U+HQ>PCU>BF>U+DCD. Different new materials showed the best effect obtained from the urea+nitrification inhibitor (U+DCD), urea+nitrification inhibitor+urease inhibitor (U+DCD+HQ), is not ideal, which were just one-fourth of the effects of U+DCD. The reason may be that, on the one hand, urea+nitrification inhibitor+urease inhibitor (U+DCD+HQ) may have a potential interaction, and on the other hand, it also has a great influence on the yield of plants. The plots treated with urea+nitrification inhibitor+ urease inhibitor (U+DCD+HQ) had the lowest yield. From a comprehensive comparison of different new material handling costs and potential environmental benefits, this study thinks that using polymer-coated urea (PCU) and biochar fertilizer (BF) has higher feasibility for the promotion of demonstration area and can achieve the desired effect. The GHGI of these two treatments were 0.099 and 0.080 kg CO_2 eq kg^{-1}, which had a comparative advantage than other treatments.

(4) Effects of different new materials for carbon sequestration and emission reduction on winter wheat yield and yield components

Different new carbon sequestration materials had different effects on yield and yield components, which had a larger effect on yield but a smaller effect on 1,000-grain weight. Among the three yield factors, grain number per spike had the largest effect on yield, among which the nitrification inhibitor and urease inhibitor overapplication (U+DCD+HQ) will result in a certain degree of yield reduction. Nitrification inhibitors (DCD) and biochar fertilizer (BF) were applied alone to increase wheat yield, among which biological carbon treatment increased wheat yield by 7.7% compared with control, and double inhibitors reduced the wheat yield by 23.2%. Considering different nitrogen fertilizer regulation measures on soil nitrogen transformation, greenhouse gas emissions and the influence of wheat yield, urea+nitrification inhibitors (U+DCD) and polymer-coated urea (PCU) compared with conventional fertilization (U) measures have obvious advantages, and in the future practical production in this area, we can try using urea+nitrification inhibitors (U+DCD) and polymer-coated urea (PCU) to realize the coordinated development of economic benefits and farmland ecological construction.

8.1.2 *Screening and Demonstration of New Models for Carbon Sequestration and Emission Reduction in Farmland*

Henan is the largest grain producer in China. In 2015, the sown area of wheat in Henan reached 5.42 million ha, with the total output exceeding 35 billion kg for the first time. Maize yield is 18.537 million tons, and wheat–maize double cropping is the main mode of grain production in most areas of Henan Province. Maize, soybean, and peanut are three common autumn crops in Huang-huai region. Although N_2O emission from farmland in this region has been studied to some extent, the effect of crop rotation on N_2O emission from farmland in this region has not been studied yet. This project aimed at wheat–maize, wheat–soybeans, wheat–peanuts rotations and other three rotation modes (Figure 8.2). And research has been carried out on greenhouse gas emissions under different rotations, aiming to build a solid carbon reduction for the region, high yield, resource efficient system of conservation tillage technology, and to provide a theoretical basis and technical support for CSA popularization.

Figure 8.2.   Test of a new model of carbon sequestration and emission reduction in wheat–maize production system.

The experiment was carried out in Ye County, Henan Province, from September 2015 to September 2018 (Figure 8.3). The contents of organic matter in the surface soil of 0–20 cm and 20–40 cm in the test site were 8.34 g kg^{-1} and 5.88 g kg^{-1}, respectively. The total nitrogen content was 961.26 g kg^{-1} and 724.80 g kg^{-1}, respectively. The content of available phosphorus and available potassium in the soil layer of 0–20 cm were 26.45 and 168.03 mg kg^{-1}. During the experiment, three treatments, namely wheat–maize, wheat–soybean, and wheat–peanut rotation modes, were set up by random block design. Each treatment was repeated three times, and there were nine plots in total, 667 m^2 for each plot. In both wheat and maize seasons, nitrogen fertilizer was applied at 225 kg N ha^{-1}, and the basal fertilizer ratio was 5:5. Soybean nitrogen fertilizer was applied at 70 kg N ha^{-1}, and basal fertilizer was applied once. The nitrogen fertilizer applied to peanut was 140 kg N ha^{-1}, and the basal fertilizer ratio was 6:4. All the crops had the same amount of P and K fertilizers, which were applied as base fertilizer once, and the amount was, respectively, 46 kg P ha^{-1} and 50 kg K ha^{-1}. Irrigation was carried out twice throughout the rotation cycle, and other management measures were the same as local conventional management measures.

Figure 8.3.   New pattern of carbon sequestration and emission reduction in wheat–maize production system.

(1) Impacts of different crop rotation patterns on greenhouse gas emissions

The N_2O and CO_2 emissions always showed higher in the wheat–maize rotation mode, compared with wheat–soybean and wheat–peanut rotation modes. However, soybean and peanut showed an emission effect on CH_4, while maize showed absorption effect on CH_4. However, the total greenhouse gas emissions of wheat–soybean and wheat–peanut cropping patterns were lower than those of wheat–maize cropping patterns. Therefore, both wheat–soybean and wheat–peanut cropping patterns can be used as climate-smart alternative cropping patterns for wheat–maize rotation in project areas.

(i) N_2O emission. Soil N_2O emission was significantly different between the two rotation cycles. In 2016, it was mainly concentrated in the summer maize season, and the peak value of N_2O emission under maize treatment reached 9,000 μg N m^{-2} h^{-1}. Peak emissions in 2017 were concentrated in the wheat sowing period, with a peak of 430 μg N m^{-2} h^{-1}, only 20th of the peak in the first year. This may be due to the fact that crop fertilization in the first summer was often accompanied by precipitation, resulting in higher soil moisture content and large amounts of nitrogen released by the denitrification process. The N_2O emission fluxes of the three rotation modes in 2018 mainly occurred from June to August in summer. The N_2O emission was mainly driven by the combined action of fertilization, rainfall, and temperature, and the N_2O emission fluxes in different planting modes were: wheat–maize>wheat–soybean>wheat–peanut.

(ii) CO_2 emission. In 2016 and 2017, the seasonal dynamic changes of soil CO_2 emissions in wheat season were basically the same, which were similar to the seasonal dynamic changes of soil temperature. In winter, the temperature was low, wheat was in the overwintering stage, the growth was slow or even stopped, the respiration of plants was weakened, and the respiration of CO_2 emissions was low with little change. After the middle of March, with the rise of temperature, wheat entered the turning green period, with vigorous growth, and wheat root respiration intensity gradually increased. However, in the growing season of maize, soybean, and peanut, there was a big difference in soil CO_2 emissions. The annual emission peak of the second rotation was only 430 mg C m^{-2} h^{-1}, which was much lower than the 1,600 mg C m^{-2} h^{-1} of the first rotation. This is due to the large ground disturbance to the test sample caused by hooded plants sampling during the first year of sampling, so the overall unhooded plants in the second year led to a significant decrease in the CO_2 flux in the second year. The CO_2 emission flux of summer crops in 2018 is consistent with the trend of 2016.

(iii) CH_4 emission. In the two rotation cycles, CH_4 showed net atmospheric CH_4 exchange entirely, especially in wheat sowing period, wheat pulling period, and maize tasseling period, with the absorption peak reaching 0.05 mg C m^{-2} h^{-1}. This is mainly related to the climatic conditions in these periods, with high temperature and no obvious precipitation. The dry and breathable soil environment is conducive to enhancing the activity of CH_4 oxidizing bacteria. The methane emission rates were different under the three rotation modes. In the wheat season, absorption was the main performance, while in the summer with high temperature and rain, the peanut and soybean fields showed an emission trend and the maize field still maintained the absorption trend.

(iv) Net effect of GHGs and emission intensity. Through 3 years of field observation, we found that the annual emissions of CH_4 and N_2O in wheat–maize rotation model were 1.25–1.06 kg C ha^{-1} and 8.59–14. 96 kg N ha^{-1}, respectively. Its global warming potential is 3,986–6,964 CO_2 equivalent, and its greenhouse gas emission intensity is 0.22–0.45 CO_2-eq kg^{-1} ha^{-1}. The total CH_4 and N_2O greenhouse gas emissions of wheat–soybean and wheat–peanut planting modes are both smaller than those of wheat–maize planting modes. The global warming potential of the wheat–peanut and wheat–soybean planting patterns is low, which may be mainly due to the fact that soybean and peanut are nitrogen-fixing crops and the planting process does not require excessive fertilization, thus greatly reducing the N_2O emission caused by chemical fertilizer. Both wheat–soybean and wheat–peanut can be used as climate-smart alternatives to wheat–maize rotation in project areas.

8.1.3 *Demonstration of Supporting Cultivation Techniques for Conservation Tillage*

The project area in Ye County aimed to establish a conservation tillage technology system featuring with carbon sequestration, emission reduction, high crop yield, and high resource efficiency, providing a theoretical basis and technical support for the promotion of CSA by setting up a two-factor experiment consisting of three tillage methods and three nitrogen application rates in this area (Figure 8.4). The experiment was carried out in Niuduzhuang Village, Longquan Township, Ye County. The organic matter contents in the surface soil layer 0–20 cm and 20–40 cm in the test site were 8.34 g kg^{-1} and 5.88 g kg^{-1}, respectively. And the total nitrogen contents were 961.26 g kg^{-1} and 724.80 g kg^{-1}, respectively. The available phosphorus and available potassium contents in the soil layer 0–20 cm were 26.45 mg kg^{-1} and 168.03 mg kg^{-1}, respectively. Farming methods and nitrogen application rate two-factor treatment, using split plot experiment, season in the maize straw returning full amount pieces, on the basis of a farming area, conventional tillage (CT), rotary tillage (RT), and no-tillage (T) were the methods used for cultivation. Deputy district for nitrogen application processing, set no nitrogen (NN, 0 kg N·ha^{-1}), low nitrogen (LN, 120 kg N·ha^{-1}), conventional nitrogen (CN, 225 kg N·ha^{-1}) were adopted as well. There were nine treatment combinations and

Figure 8.4. Demonstration of protective cultivation technology for wheat–maize production system.

the area of each plot was NN (10 m × 18 m), LN and CN (50 m × 18 m). The amount of phosphorus fertilizer (P_2O_5) and potassium fertilizer (K_2O) used was 150 kg·ha^{-1}, which was used as the base fertilizer before the tillage operation. Urea was applied in the jointing stage of wheat, and the base topdressing ratio of nitrogen was 5:5. Other management measures were the same as other high-yielding fields.

(1) Effects of different tillage methods on soil enzyme activity
Tillage methods had a great influence on soil enzyme activity, among which CT was beneficial to increase the activity of soil sucrase and urease. The soil sucrase activity was increased by increasing the nitrogen application, but it had no significant effect on urease activity. In the 0–20 cm soil layer, the activity of soil sucrase decreased slowly from the turning green stage to the filling stage of winter wheat, and increased suddenly in the maturing stage. Under the condition of low nitrogen (LN), the sucrase activity in the 0–20 cm soil layer under the conventional tillage (CT) measures was stronger. The activity of soil sucrase in no-tillage (NT)

treatment of 0–20 cm soil layer could be improved to some extent by increasing nitrogen application. The effects of cultivation and nitrogen application on the activity of sucrase were significant. The effect of different nitrogen applications on urease activity was not obvious, and the effect of different tillage methods on urease activity was prominent. Under the condition of low nitrogen (LN), the conventional tillage (CT) method was beneficial for improving urease activity. The effects of tillage and nitrogen application on urease activity were significant. The urease activity in the 20–40 cm soil layer was significantly lower than that in the 0–20 cm soil layer. Among the three tillage methods, the urease activity of the 20–40 cm soil layer at the turning green stage was stronger under NT conditions. Under the condition of CT, the urease activity in 20–40 cm soil layer was stable. Before the jointing stage of wheat, nitrogen application had a great influence on urease activity in 20–40 cm soil layer. After the flowering of wheat, the effect of nitrogen application on urease activity in 20–40 cm soil layer was not obvious compared with the early stage.

(2) Impacts of different farming practices on greenhouse gas emissions

No tillage (NT) treatment can effectively reduce greenhouse gas emissions compared with conventional tillage (CT). With the increase of nitrogen fertilizer use, N_2O and CO_2 emissions increased, but CH_4 absorption also increased. No tillage+conventional nitrogen application (NT+CN) and conventional tillage+low nitrogen (CT+LN) provided the best reduction effect.

(i) N_2O emission. After sowing winter wheat, N_2O emission peaked. Between sowing and seedling stage, under the treatment of no-tillage (NT) and rotary tillage (RT), the N_2O emission gradually slowed down, while the tillage (CT) treatment showed a steady rise. However, the N_2O emission flux under the treatment of the three tillage methods in the overwintering stage all decreased rapidly, and the N_2O emission flux showed an upward trend after topdressing in the jointing stage, with the peak appearing from the jointing stage to the grout stage. At a low nitrogen application level, the N_2O emission flux under showed as: RT>CT>NT, and at a high nitrogen application level, it was CT>RT>NT. Under the same tillage method, increasing nitrogen application increased N_2O emission in wheat field.

(ii) CO_2 emission. The experimental gas collection chamber was a dark chamber, and the CO_2 flux represented the total respiratory flux of the ecosystem, including soil respiration, root respiration, and plant respiration. The test results in 2016 showed that the peak of CO_2 emission in wheat season occurred between jointing period and filling period after topdressing irrigation. Under different nitrogen application levels, the CO_2 emissions of wheat season were all NT>RT>CT. Under the same tillage method, there was an obvious positive correlation between CO_2 emission and nitrogen application, and reducing nitrogen application was helpful to reduce CO_2 emission of wheat field to a certain extent. In the 2018 wheat season, there were two CO_2 emission peaks, which happened after wheat planting and after fertilization and irrigation at jointing stage. Under no nitrogen (NN) treatment, the CO_2 emission flux of the whole growth period of no-tillage (NT) was lower than that of rotary tillage (RT) and conventional tillage (CT). Among these, the CO_2 emission flux of rotary tillage (RT) before jointing period was higher than that of conventional tillage (CT), while that of conventional tillage (CT) after jointing period was higher than that of rotary tillage (RT). Under low nitrogen (LN) treatment, the CO_2 emission flux of no-tillage (NT) treatment was smaller than that of rotary tillage (RT) and conventional tillage (CT), and the performance between rotary tillage (RT) and conventional tillage (CT) was consistent with that of no nitrogen (NN). Under the conventional nitrogen (CN) treatment, the CO_2 emission fluxes during the whole wheat growth period followed the trend: NT < RT < CT. Under the same tillage method, there was a positive correlation between nitrogen application and CO_2 emission flux.

(iii) CH_4 emission. The wheat field showed an overall absorption of CH_4, and the experimental results in 2016 showed that the absorption peak occurred from the jointing stage to the flowering stage of wheat. The uptake of CH_4 by no-tillage (NT) and rotary tillage (RT) was higher under no nitrogen (NN) or low nitrogen (LN) treatments. The CH_4 uptake in wheat fields was higher than that in low nitrogen (LN) treatment or no nitrogen (NN) treatment under conventional nitrogen (CN) treatment. The observation results showed that the increase of nitrogen application could promote the absorption of CH_4 gas in wheat fields to a certain extent, especially under the condition of conventional tillage (CT). There was an obvious interaction effect between nitrogen application and tillage on CH_4 emission. The 2018 test results showed that the two absorption peaks of CH_4 in wheat season occurred after wheat planting and after fertilization and irrigation at jointing stage. Under the condition of low nitrogen (LN), the absorption of CH_4 by no-tillage (NT) was significantly stronger than that by rotary tillage

(RT) and conventional tillage (CT) during the whole growth period of wheat. Under the condition of conventional nitrogen application (CN), the methane absorption intensity of no-tillage (NT) treatment was higher than that of rotary tillage (RT) and conventional tillage (CT). Except that conventional tillage (CT) was higher than that of rotary tillage (RT) at the jointing stage before topdressing, rotary tillage (RT) was stronger than that of conventional tillage (CT) at other stages. At the level of conventional nitrogen application (CN), no tillage (NT) treatment had higher CH_4 absorption intensity than rotary tillage (RT) and conventional tillage (CT) treatment, but there was no significant difference in CH_4 absorption between rotary tillage (RT) and conventional tillage (CT) treatment. Under the same tillage method, the absorption intensity of CH_4 in wheat fields showed an increasing trend with the increase of nitrogen application.

(iv) Comprehensive greenhouse gas emissions. The test results in 2016 showed that under the no-tillage (NT) mode, the cumulative CO_2 emission in wheat season under the conventional nitrogen (CN) treatment was the largest, which was 29,279 kg C ha^{-1}. Under the rotational tillage (RT) treatment, the maximum CH_4 uptake accumulation was 3.2 kg C ha^{-1} in the wheat season under the no nitrogen (NN) treatment, and the maximum N_2O emission accumulation was 8.0 kg N ha^{-1} in the wheat season under the no-tillage+low nitrogen (NT+LN) treatment. Under different tillage treatments, the emission of CO_2 and N_2O was significantly reduced by reducing nitrogen application, but the absorption of CH_4 was increased. The effects of different tillage methods on the accumulation of CO_2 and N_2O emissions were as follows: NT>RT>CT, while the effects on CH_4 showed an opposite trend. There were obvious interaction effects between nitrogen application and tillage. The 2018 test results showed that greenhouse gas emissions, especially N_2O and CO_2 emissions, increased with the increase of nitrogen application. The effects within different farming methods are as follows CT>RT>NT, and the global warming potential is consistent with the greenhouse gas emission trend. In terms of greenhouse gas emissions per unit grain yield, no-till+conventional nitrogen application (NT+CN) and conventional tillage+low nitrogen application (CT+LN) performed the best, both with 1.50 kg CO_2-eq kg^{-1}. Normal nitrogen application also showed better balance between carbon emissions and yield among different tillage methods. The results showed that there was a significant interaction between nitrogen application and tillage in terms of the greenhouse gas emission flux. Under different N application levels, the N_2O emission flux of wheat fields under rotary tillage (RT) was significantly higher than that of the other two tillage methods. From the overall trend, increasing nitrogen application increased N_2O emission in wheat fields. Reducing nitrogen application can effectively reduce greenhouse gas emissions from wheat fields.

(3) Effects of different tillage methods and nitrogen application rates on soil net nitrogen mineralization rate

It was found that the net mineralization rate of soil nitrogen was positively correlated with three types of nitrogen application. The net mineralization rate of soil nitrogen had a tendency to decrease with the decrease of nitrogen application. Under rotary tillage (RT) and no tillage (NT), the net nitrogen mineralization rate of soil was higher, which was especially obvious in the later stage of wheat growth.

Nitrogen mineralization rate was very low between wheat turning green and jointing stages, and net nitrogen mineralization rate of soil from jointing to flowering was significantly higher than that of other periods, and nitrogen mineralization rate of soil from flowering to maturity decreased significantly after flowering. This indicated that the mineralization of soil nitrogen by microorganisms is most intense from jointing stage to flowering stage, which was closely related to the increase of temperature and sufficient rainfall during this period. On the whole, the net nitrogen mineralization rate of soil under no-tillage (NT) was significantly higher than that under rotary tillage (RT) and conventional tillage (CT). The nitrogen mineralization rate of both no-tillage (NT) and conventional tillage (CT) was LN>CN>NN, while that under rotary tillage (RT) was CN>LN>NN.

(4) Effects of interaction between tillage practices and nitrogen management on emergence and yield of wheat

The experimental results showed that the seedling emergence rates of wheat under different tillage methods under three nitrogen fertilizer treatments were as follows: CT>RT>NT, and there were significant differences among different tillage methods, among which the best seedling emergence was conventional tillage+conventional nitrogen application (CT+CN) treatment. Under the same tillage method, seedling emergence rate was similar between different application amounts, and the conventional nitrogen (CN) treatment was slightly higher than the other two treatments. The results showed that the physical characters of the tillage layer caused by tillage have a great influence on the emergence rate of wheat, while the effect of nitrogen application is relatively small.

Under the same nitrogen application rate, except for the conventional nitrogen application (CN), no-tillage (NT) was slightly less than rotational tillage (RT). The overall performance was NT>RT>CT. The 1,000-grain weight was not significantly different among treatments. Under the same tillage method, the wheat yields shown as CN>LN>NN. Under the condition of conventional nitrogen application (CN), the yield was the highest in rotary tillage (RT). Conventional tillage (CT) treatment had the highest yield under both conventional nitrogen (CN) and low nitrogen (LN) conditions. On the whole, according to the local customs and practices of conventional nitrogen+no tillage (CN+NT), the yield of wheat sown with no tillage (NT) was slightly higher than that of rotational tillage (RT) but lower than that of conventional tillage (CT) after the total amount of maize stalk was returned to the field. There is still a great potential to increase yield in the local farmland in terms of fertilization management.

8.2 Climate-smart Wheat–Maize Production Technology Training and Consultation Service

Technical training and consultation service are the important platforms to impart new technologies and new ideas to front-line producers, especially to strengthen farmers' (or agricultural producers and operators) understanding and adoption of CSA. China's CSA projects in Ye County Henan Province were implemented by conducting technical training and consulting service to the project area farmers and agricultural production operator, grassroots management personnel, technical promotion personnel to introduce the concept of intelligent agricultural climate, policy, and technology (Figure 8.5). Mainly training sessions aimed at climate-smart related research and practice of agriculture and the key technology of wheat–maize planting patterns. During the period of 2016–2019, the project team carried out a number of technical training courses, field technical guidance, and thematic discussions, as well as organized and compiled relevant technical specifications and operational specifications, which have effectively promoted the implementation of China's climate-smart agriculture project and provided a large number of materials for further promoting CSA.

Projects focused on the demonstration area of wheat–maize production system in Ye County, Henan Province, provided a detailed

Figure 8.5.   Farmer training in wheat–maize production system.

introduction on the key to the carbon emission increase in crop production technology integration and demonstration, the innovation and application of supporting policies, and promotion and development of public knowledge and other activities. The project team conducted training 28 times between 2016 and 2019. There were more than 40 cumulative preparation training courseware and video materials, 50 copies of materials related to technical training on production and service, and more than 13,000 guides for training farmers. Through training and guidance, the project team hoped to improve the utilization efficiency of inputs such as chemical fertilizers, pesticides, and irrigation water and the working efficiency of agricultural machinery to reduce the carbon emission of crop system and to increase the carbon storage of farmland soil. Through technical training and service, the system of climate-smart crop production should be established, the adaptability of crop production in the project area to climate change should be enhanced, energy conservation and emission reduction of agricultural production should be promoted, and theoretical guidance and technical support should be provided for wheat and maize production in this county to cope with climate change.

8.2.1 *CSA Concepts and Relevant Technical Training*

From 2016 to 2019, relevant expert teams were organized to carry out training on the concept and model of CSA development in the project demonstration area several times, teaching core issues such as what is CSA, how to develop it, and China's CSA development model. The significance of implementing CSA was publicized through videos and print material, and agricultural technicians and large grain growers were guided to visit CSA experimental facilities, and the analysis of the effect of energy conservation and emission reduction was explained. Through training, the project team greatly popularized the concept of CSA, and the participants included cooperatives and grain investors. These training activities helped participants understand the concepts of solid carbon, energy conservation and emissions reduction, adaptation to climate change; study the ability to increase crop production to adapt to climate change and other technology, and through WeChat group, symposiums were conducted fo other farmers in the demonstration area. At the same time, circuit training strengthened the farmers cognition in demonstration area and promoted the local agricultural sustainable development.

8.2.2 *Technical Training and Guidance for High Yield and Efficient Crop Management*

From 2016 to 2019, the project training team organized relevant experts to carry out activities many times, such as classroom training, on-site consultation, and field technical guidance in Quanyin Village, Loufan Village, Ximuzhuang Village, and Guolizhuang Village. They synthetically organized more than 20 special training sessions, such as the "Winter Wheat Field Management Technology", "Wheat Seedling Stage Difference and its Reason Analysis", "Wheat Quality Efficiency Cultivation Technique", "Wheat Soil Testing Formula Fertilization Technology", "Wheat High Yield and Fine Chemical Regulation Technology", "Efficient Fertilization Technique for Wheat", "High Quality Wheat Seeding Technology", "Wheat Seed Treatment before Sowing and Seed Dressing Agent Application Technology", "Wheat Seed Treatment Agent", "High Yield and High-Efficiency Technology Training for Wheat", "Adjustment of Maize Planting Industry structure", "Maize Production Situation", "High Yield and Efficient Cultivation Technique of Maize", "Maize Production Problems and Key Cultivation Techniques in Henan Province", "Overview of Soybean Production in Henan Province and Introduction of New Soybean Varieties", "High-yield Peanut Cultivation Techniques", "High-yield Soil Cultivation and Soil Fertility Enhancement Technology", and "The Basic Meaning and Principle of Fertilizing Fertilizer".

The training content included high-quality seeds such as the selection of wheat sowing date and quantity, land preparation and fertilization before sowing, seed dressing, pest and disease control, selection and proper use of chemical fertilizers and pesticides, and mechanized management, emission reduction and efficiency enhancing cultivation techniques, quality improvement, and harvest management measures. Training was also given on the selection of maize varieties, fertilizer selection and application, disease and pest control in middle and late stages, disaster prevention and mitigation, and late harvest. At the same time, according to the current situation of peanut planting in Henan Province, the peanut production and management were explained in detail from the aspects of variety selection, disease and pest control, appropriate harvest, drug damage and prevention, and peanut mechanization.

Through training, large grain growers, cooperatives, and farmers further mastered the high yield and high-efficiency cultivation technology of crops, the technology of fertilizer optimal management, peanut and soybean high yield and high-efficiency cultivation technology, and realizing the high yield and high efficiency cultivation of peanut and soybean. Expanding cultivated area of wheat–soybean and wheat–peanuts enhanced the tolerance of crop production to deal with extreme weather events, improved the efficiency and effectiveness of the demonstration area of food production, motivated the local planting structure adjustment, alleviated the widespread problem of high investment, low utilization rate, reduced the greenhouse gas emissions, and promoted the environmentally friendly agricultural sustainable development.

8.2.3 *Conservation Tillage Techniques for Crop Production*

From 2016 to 2019, the consulting team of Ye County project area had organized many well-known experts in related fields to carry out classroom training and field technical guidance centered on the theory and practice of conservation tillage technology, and they organized several training sessions such as the "Conservation Tillage", "Conservation Tillage Machines and Tools and Technical Specifications of Henan Province", and "Mechanization Production Technology of

Huang-huai-hai Soybean with No-tillage Mulching". These sessions systematically introduced the theoretical knowledge and typical cases of conservation tillage, protective tillage equipment for wheat and maize, conservation tillage types and their technical specifications in Henan Province, the mechanization production technology of soybean stubble with no-tillage, and so on. Through training, large grain growers and farmers further mastered the technology of less tillage and no-tillage and crop straw mulching, which effectively promoted the application of wheat–maize production system straw returning and conservation tillage, and promoted the popularization of climate-smart wheat–maize production technology.

8.2.4 *Green-Integrated Crop Control Technology*

From 2016 to 2019, the consulting team of Ye County project area organized several well-known experts to carry out training and practical guidance on green pest prevention and control technologies in the project area. The training area included Quanyin Village, Loufan Village, Ximuzhuang Village, Guolizhuang Village, Dahezhuang Village, and Caozhuang Village (Figure 8.6). The training content consisted of more than 10 sessions, such as the "Integrated Pest and Disease Control in Wheat", "Pest and Disease Control Technology in Late Growth Stage of Wheat", "Weeds and Chemical Control in Wheat Fields", "Identification and Control of Major Wheat Diseases and Insect Pests", "Wheat Disease Identification, Diagnosis and Comprehensive Prevention and Control Technologies", "Identification, Diagnosis and Prevention and Control Techniques for Maize Diseases and Insect Pests", "Integrated Solutions for Wheat Pests and Diseases", and "Identification and Control of Weeds in Wheat Fields" and field technical advisory guidance. The training content covered the main species of weeds in wheat fields of Henan Province, the harm of weeds in wheat fields, occurrence and control technology of wheat diseases, occurrence regularity and optimal control period of weeds in wheat field, a brief introduction of herbicides commonly used in wheat fields, problems to be paid attention to when using herbicides in wheat fields and remedial measures for herbicides in wheat fields, and the regularity of occurrence and identification and control technology of maize diseases and insects. As a result, large grain growers and farmers further mastered the efficient pest control technology and pesticide reduction and

Figure 8.6.   Wheat–maize production system demonstration area training.

application technology, which greatly promoted the green and high-quality development of local agriculture.

8.3 Monitoring and Evaluation of Climate-Smart Wheat–Maize Production

To get timely feedback, adjust the project execution strategy, and track the overall effect of the project, an important strategy is to establish an appropriate dynamic monitoring and evaluation mechanism. By comparing with the set baseline, we can timely and accurately evaluate the progress of CSA projects, reflect on the problems existing in project implementation and propose corresponding countermeasures, which can effectively ensure the progress of the project and achieve the set goals. During the 2016–2019 project period, we focused on tracking the crop production, carbon sequestration, and emission reduction, environmental effects, pest control effects, and social impacts in the project area.

Since the implementation of the project in 2016, the results of monitoring and farm household surveys showed that (1) the yield of the wheat–maize crop production system in the project area has shown an upward trend year on year, while the annual yield was higher than the baseline yield and (2) the implementation of the project significantly reduced farmland N_2O emissions and increased soil organic carbon content. Compared with the baseline, the wheat and maize seasons reduced emissions by more than 50% and 20%, respectively, and increased organic carbon storage by more than 1%; (3) The forest carbon sink of the wheat–maize production system continued to increase; (4) Surface water and groundwater pollution were reduced in the project area; (5) The frequencies of diseases, pests, and weeds in the wheat–maize production system in the project area were lower than the baseline. At the same time, the frequency of pesticide use and the cost of pesticides in the project area were lower than those in the non-project area; (6) Using large wide-width pesticide applicator and knapsack electrostatic sprayer improved the utilization rate of pesticides and reduced the cost of pesticides. Overall, the implementation of the climate-smart project for the wheat–maize production system has enhanced the stability of the yield in the project area, reduced agricultural production costs, and promoted farmers' income. At the same time, the reduction in the use of pesticides and fertilizers has strengthened farmers' climate-smart concept, which is conducive to energy conservation and emission reduction in project areas. The combination of climate intelligence projects with national strategies such as rural revitalization and targeted poverty alleviation has played a role in driving disadvantaged groups and helping to alleviate poverty.

The monitoring and evaluation process showed in detail the achievements of the project in stages, and effectively reflected the technical effects and related results of the key measures of the project. By summarizing the project implementation process, it provided guidance for further development of China's CSA.

8.3.1 *Monitoring and Evaluation of Carbon Sequestration and Emission Reduction*

The project team used the winter wheat-summer maize cropping system in Ye County as the research object to determine the project area, project boundary, project target group, and project activity objects. According to the application of the relevant implementation technology of the project, we carried out the design of the monitoring and evaluation plan for the project area, selected sample villages, sample farmers and sample farmland, and deployed monitoring points and monitoring areas in a targeted manner (Figure 8.7). Based on field observations and DNDC models, we studied and predicted the overall carbon sequestration and emission reduction capabilities of the project area and optimized the improvement

Figure 8.7.   Carbon sequestration and emission reduction monitoring of wheat–maize cropping system.

methods. We also quantified the carbon sequestration and emission reduction potential of CSA to provide scientific support for its promotion.

Since 2015, monitoring points have been arranged in Ye County, Henan Province, and experimental monitoring has been initiated. According to the requirements of the project and the needs of technical research, we systematically monitored the changes in soil carbon sinks, greenhouse gas emissions, yield composition, and other aspects of the winter wheat-summer maize cropping system at this test site. We strive to comprehensively evaluate the carbon sink emission reduction effect of the project.

(1) Dynamic monitoring of crop yield

During the implementation of the project, the project team selected 124 yield monitoring sample points in the wheat–maize cropping system in Ye County, Henan Province, for yield information collection (Figure 8.8), of which five replicate samples were taken from each sampling point.

In the 2016 project area, the winter wheat yield under the project conditions was lower than the yield under the baseline conditions. In 2016, under the baseline conditions of the Ye County project area, the winter wheat yield was 6,031 kg ha^{-1}, and under the project conditions, the winter wheat yield was 5,161 kg ha^{-1}, which was 14.4% lower than the baseline conditions; In 2017, the winter wheat yield under the project conditions was higher than that under the baseline conditions. In 2017, the winter wheat yield under the project conditions was 7,272 kg ha^{-1}, and the yield under the baseline conditions was 6,715 kg ha^{-1}. Compared with the baseline conditions, the project conditions increased the yield by 8.3%. In 2018, the yield of winter wheat under the project conditions

Figure 8.8.   Farmland and forest protection net in wheat–maize cropping system.

compared with the baseline conditions showed an increasing trend. Under the project conditions, the yield of winter wheat in Ye County was 8,125 kg ha⁻¹, which was 6.5% higher than the baseline conditions. The yield of winter wheat under the project conditions in the first year was lower than that under the baseline conditions. The main reason was that agricultural management measures caused greater disturbance to soil and crop growth, which would have a greater impact on yield or even decrease yield. However, after the second year of project implementation, due to the reduction of interference from various measures on winter wheat, the yield of winter wheat under project conditions has increased, and the project's yield increase or stabilization effect began to gradually appear, achieving the expected results.

All the summer-maize yields in the project area showed that the yield under the project conditions was higher than the yield under the baseline conditions. In 2016, the summer maize yield under the project conditions was 5,549 kg ha⁻¹, and under the baseline conditions was 4,736 kg ha⁻¹. Compared with the baseline conditions, the yield increase rate under the project conditions is 17.2%. In 2017, under the project conditions, the summer maize yield was 9,228 kg ha⁻¹, the baseline conditions yield was 8,724 kg ha⁻¹, and the yield increase rate under the project conditions was 5.8%. In 2018, the yield of summer maize in the project area was 8,972 kg ha⁻¹ under project conditions and 8,612 kg ha⁻¹ under baseline conditions. Compared with the baseline conditions, the yield of project conditions increased by 4.2%. According to the results of the 3-year experiments in 2016–2018, intervention measures under project conditions can increase the yield of summer maize.

(2) Characteristics of greenhouse gas emissions from farmland
In the project area, the N_2O emissions in 2016 and 2017 were both lower under project conditions than under baseline conditions. In 2016, the N_2O emission of winter wheat under the project conditions was 0.59 kg N ha⁻¹, and the emission under the baseline conditions was 1.22 kg N ha⁻¹. Under the project conditions, the N_2O emission was reduced by 51.64%. In 2017, the N_2O emissions under the winter wheat project conditions were 0.47 kg N hm⁻², which was 54.37% less than the 1.03 kg N ha⁻¹ emissions under the baseline conditions. It showed that project treatment measures in this area can effectively reduce N_2O emissions.

In the summer maize season, the N_2O emissions from 2016 to 2018 were all lower under the project conditions than under the baseline conditions. In 2016, the N_2O emission of winter wheat under the project conditions in the project area was 0.59 kg N ha^{-1}, and the emission under the baseline conditions was 1.22 kg N ha^{-1}. The N_2O emission under the project conditions was reduced by 51.64%. In 2017, the N_2O emission under the conditions of the summer maize project was 0.47 kg N ha^{-1}, which was 54.37% less than the 1.03 kg N ha^{-1} emission under the baseline conditions. In 2018, summer maize under project conditions reduced N_2O emissions by 20.96% compared to summer maize under baseline conditions. These showed that the implementation of the project can effectively reduce N_2O emissions in the summer maize season in this area.

(3) Fuel consumption in tillage and fertilization

After investigating farmers in the project area of Ye County, we found that irrigation mainly uses well water, and some fields near rivers are irrigated by river water. Due to the lack of a supporting high-voltage power system, the irrigation uses diesel pumping combined with common pipelines. In general, wheat and maize are irrigated once each, and the average diesel consumption is 96 L ha^{-1} each time. Then we use the formula to calculate the fuel consumption emission (882.2 t CO_2). Since this data is not accurately observed based on the conservative calculation principle of greenhouse gas, it was not included in the calculation of total emission reductions.

(4) Soil carbon storage

The soil carbon storage in the project area is based on the calculation method in *CSA-C-3 — Monitoring and Evaluation of Carbon Sink Emission Reduction*. The calculation results showed that during 2016–2018, the soil carbon storage under the technical conditions of the project area likely increased. In other words, intervention measures under project conditions can enhance the soil's carbon sink function. It showed that with the continuous implementation of project intervention measures, the carbon sink function of the soil in this area could be enhanced.

The soil type in the project area is mainly yellow cinnamon soil, with low soil nutrients but with good fertilizer retention performance. In 2019, the organic matter contents of the top 0–30 cm soil layer in the project area varied from 19.88 to 21.61 g kg^{-1}; The soil organic carbon content per unit area was 46.03±1.05 t C ha^{-1}, which increased by 13.62% compared to the baseline. The demonstration area of carbon sequestration and emission reduction technology in Ye County is 3,480.9 hectares. Together with 2018 reports, we could conclude that the 2019 soil carbon storage of top 0–30 cm soil layer in the project area increased by 25,982 t CO_2eq compared with 2018 (Table 8.1).

(5) Forest carbon sink

From 2016 to 2018, the forest carbon sink in the project area has shown an upward trend year on year with the continuous advancement of the construction of farmland forest protection net. In 2016, the number of trees planted in the project area of Ye County was calculated at 7,000 trees per year, and the carbon sequestration amount was 143.94 t C. In 2017, more than 20,000 seedlings of various types such as paulownia were planted in the Ye County project area. According to the formula calculation, the carbon sequestration of forest trees in Ye County in 2017 reached 920.92 t C yr^{-1}.

8.3.2 *Environmental Effect Monitoring and Evaluation*

As the international communities pay more and more attention to climate change, greenhouse gas emission reduction, and food security, the research on farmland soil carbon sequestration and emission reduction technology have received unprecedented attention from the scientific community. It is essential to promote the application of energy-saving and carbon sequestration technologies under the premise of ensuring food production in the main grain production areas and essential to conduct the demonstration and evaluation of emission reduction effects, which not only improve soil fertility and productivity, and reduce soil greenhouse gas emissions, but also provide China's strategic choices for maintaining sustainable agricultural development.

Table 8.1.   The soil carbon storage changes in the project area of Ye County, Henan Province from 2016 to 2019.

Year	Condition	DSOC(t C ha^{-1})	Area(ha)	ΔDSOC(t C ha^{-1})	ΔSOC(t CO_2)
	Baseline	40.52 ± 0.66			
2016	Project	40.75 ± 1.13	863.27	0.24	748
2017	Project	41.58 ± 1.68	1.318.60	0.83	4.017
2018	Project	44.00 ± 1.37	1.669.20	2.41	14.776
2019	Project	46.03 ± 1.05	3.480.90	2.04	25.982

Ye County has a good agricultural production base, and the land area is very large. Wheat–maize is the main cropping system in this area, and it has a good representation in the main grain production areas of Huang-huai-hai Region. The project area included 21 villages: Quanyin, Guoluzhuang, Beidaying, Niuduzhuang, Loufan, Ximuzhuang, Quanji, Tiezhang, Dahezhuang, Zhongzhang, Caozhuang, Xiaohewang, Xiaoheguo, Longquan, Jiazhuang, Baihaozhuang, Caochang, Wuzhuang, Shenzhuang, Nandaying, Dawanzhang, and Longquan Township. And there are seven administrative villages in Yeyi Town, including Caizhuang, Wandukou, Sicheng, Duanzhuang, Shenwan, Lianwan, and Tongxinzhai A total of 28 administrative villages in two townships serve as demonstration areas for the wheat–maize dual cropping system.

The project began to be implemented in 2015. With the implementation of various programs of the project, including soil testing and formula fertilization, formula deep fertilization, crop reducing pesticide application, straw return, land leveling, the application of chemical fertilizers and pesticides can be reduced. The reduction of backwater reduces the agricultural pollution sources entering the waters of the project area and reduces the risk of harmful substances and nutritional pollutants entering the water body. It reduced the application of chemical fertilizers and pesticides, reduced farmland water return, reduced agricultural pollution sources entering the waters of the project area, and reduced the risk of harmful substances and nutritional pollutants entering the water body. Through the monitoring of surface water and groundwater in the project area, the environmental benefits are clear now.

(1) Arrangement of monitoring points in the project area
The main cropping system in the project area is wheat–maize rotation. Irrigation in the project area includes reservoir irrigation and groundwater irrigation. Water from the Gushitan irrigation area is used for reservoir irrigation. Wells that are 20–50 m deep are used for groundwater irrigation and farmland drainage enters the Li River. Sampling types in the project area of Ye County mainly includes: (i) Reservoir water diversion and irrigation: the sampling points were located at the inlet of the project area of Gushitan irrigation district; (ii) The water outlet of the project area: the sampling points were located at the outlet of the Li River project area; (iii) 20–50 meters of groundwater: the main choices were Wandukou, Shenzhuang, Tongxinzhai, and Caizhuang.

(2) Sampling monitoring content

We set up monitoring sites based on information such as irrigation type, soil classification, crop planting area and river distribution in the project area, and monitored surface water at least twice a year. The measurement indicators included CODCr, NH^+_4-N, TN, TP, and NO^-_3-N, etc.

Groundwater is the source of drinking water for both residents and the irrigation water for the project area in Henan Province. The quality of groundwater is very important to the health of local people. We set groundwater monitoring sites based on the irrigation type, soil classification, and planting area of the project area. Groundwater was monitored at least twice a year. The measurement indicators included permanganate index, NO^-_3-N, TN, TP, and NH^+_4-N.

The main evaluation methods of water quality monitoring included single-factor evaluation method, multi-factor evaluation method, and entropy weight fuzzy matter-element evaluation model. This study chose the entropy weight fuzzy matter-element model method to calculate weight in the multi-factor comprehensive evaluation. The entropy-weighted fuzzy matter-element model has great resolution for each evaluation index. The size of the weight value in the evaluation is the key factor of the evaluation, and its size will play a decisive role in the evaluation result. The basic principle of the entropy method is to determine the weight of each index according to the degree of change in the value of each index. The greater the degree of index value changes, the smaller the information entropy value will be, indicating that the greater the amount of information provided by the index, and the greater the corresponding weight is; conversely, the smaller the weight, the less the information. This method avoids the error caused by human scoring and uses the change of the factor itself to objectively obtain the weight value for evaluation.

(3) Water quality in the project area

The water quality monitoring of Ye County, Henan Province (Figure 8.9), began in 2014. Sampling and testing were conducted in January 2014, June 2015, May 2016, December 2016, April 2017, July 2017, December 2017, April 2018, November 2018, and May 2019. A total of five monitoring points were selected for monitoring in Ye County. The monitored indicators included the chemical oxygen demand (COD), ammonium content (NH^+_4-N), total phosphorus (TP), and total nitrogen (TN) content of the water body. The monitored water bodies included

Figure 8.9. Water quality monitoring of wheat–maize cropping system.

surface water and groundwater. The surface water monitoring results are evaluated according to the national standard (GB 3838-2002) for water quality categories, and the groundwater is evaluated according to the national standard (GB/T 14848-93). Overall, with the implementation of the project, the water pollutants in the groundwater and surface water of Ye County have decreased year on year, and the water quality has improved (Figure 8.10).

According to the observation results of surface water, the surface water quality in the project area was in constant fluctuation from 2014 to 2018. We observed various indicators of surface water measurement and found the COD and TN of surface water have been significantly reduced (Figure 8.11), which indicated that the implementation of the project may have the effect of reducing COD and TN in surface water. But it has no obvious effect on NH_4 and TP in surface water (Table 8.2).

The quality of groundwater in Henan Province is poor. However, by observing the previous monitoring results during the implementation of the project, it was found that the quality of groundwater was showing a trend of gradual improvement (Table 8.3). We analyzed the monitoring

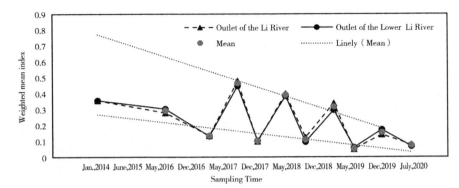

Figure 8.10. Weighted average index of surface water quality in the wheat–maize cropping system.

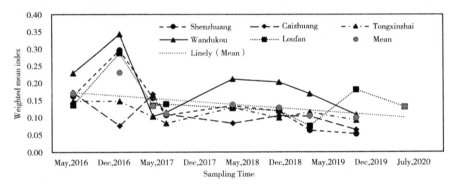

Figure 8.11. Weighted average index of groundwater quality in wheat–maize cropping system.

results of surface water and groundwater in Henan Province. The implementation of the project effectively reduced the number of pollutants in the water body and improved the water quality (Figure 8.11).

8.3.3 *Monitoring and Evaluation of Pest Management*

Given the actual situation of agricultural production in the project area, under the premise of ensuring food production, the implementation of the project reduced the use of pesticides and fertilizers, carried out integrated

Table 8.2. 2014–2019 wheat–maize cropping system surface water monitoring results (mg l^{-1}).

Monitoring date	Monitoring point	COD	NH$^+_4$-N	TP	TN	Water quality category
January 2014	Li River Outlet Section	8.49	0.09	0.04	3.93	V
January 2014	Water intake section of Gushitan Reservoir	2.29	0.13	0.07	3.23	V
June 2015	Outlet of the Lower Li River	<15.00	0.11	0.03	2.75	InferiorV
June 2015	Outlet of the Lower Li River	<15.00	0.09	0.02	3.13	InferiorV
May 2016	Outlet of the Lower Li River	<15.00	0.11	0.05	0.914	III
May 2016	Outlet of the Lower Li River	<15.00	0.08	0.04	0.894	III
December 2016	Outlet of the Lower Li River	<15.00	<0.05	0.10	5.09	Inferior V
December 2016	Outlet of the Lower Li River	<15.00	<0.05	0.06	4.81	Inferior V
April 2017	Outlet of the Lower Li River	9.90	0.07	0.02	0.753	III
April 2017	Outlet of the Lower Li River	10.60	0.09	0.03	0.753	III
December 2017	Outlet of the Lower Li River	5.40	0.05	0.03	4.64	Inferior V
December 2017	Outlet of the Lower Li River	4.50	0.05	0.03	4.51	Inferior V
May 2018	Outlet of the Lower Li River	4.00	0.11	0.04	1.20	IV
May 2018	Outlet of the Lower Li River	<2.30	0.08	0.03	0.97	III
December 2018	Outlet of the Lower Li River	2.30	0.12	3.89	0.05	Inferior V

(*Continued*)

Table 8.2.   (*Continued*)

Monitoring date	Monitoring point	COD	NH$^+_4$-N	TP	TN	Water quality category
December 2018	Outlet of the Lower Li River	2.30	0.09	3.44	0.03	Inferior V
May 2019	Outlet of the Lower Li River	3.00	0.10	0.02	0.36	II
May 2019	Outlet of the Lower Li River	3.00	0.07	0.02	0.52	III
December 2019	Outlet of the Lower Li River	5.50	<0.025	0.02	1.49	V
December 2019	Outlet of the Lower Li River	6.10	<0.025	0.02	1.84	V

pest control, controlled the degree of pest damage, and reduced pesticide pollution. Therefore, the implementation of the project should pay more attention to the application of integrated pest management technology and specialized unified prevention and control. Integrated pest management (IPM) is a strategy for controlling pest, which is based on the overall perspective of the agricultural ecosystem and based on the relationship between pests and the environment. It coordinated the use of various measures such as agriculture, physical, biological, and chemical prevention and control, gave full play to the role of natural control factors in agricultural ecology, and controlled agricultural pests below the allowable level of economic losses. IPM attaches great importance to the application of integrated control technologies including resistant varieties, cultivation measures, biological natural enemies, and chemical agents, especially the use of biological control factors such as natural enemies to control pests and diseases and adopts a cautious attitude toward the use of chemical pesticides. The specialized prevention and control of crop diseases and insect pests refers to the service organizations with corresponding plant protection expertise and equipment to carry out socialized, large-scale, and intensive crop diseases and insect pest prevention and control services. Specialized control is the product of agricultural development to a certain stage, conforms to the general laws

Table 8.3. 2014–2019 Groundwater monitoring results of wheat–maize production system (mg l^{-1}).

Monitoring date	Monitoring point	Permanganate index	TP	TN	NO$_3^-$-N	Water quality category
January 2014	Wandukou	1.58	0.13	4.83	4.395	II
	Shenzhuang	2.06	0.02	2.71	2.741	III
	Tongxinzhai	2.32	0.21	2.16	2.203	III
	Caizhuang	2.51			4.533	III
June 2015	Shenzhuang	0.51	<0.01	6.38	9.75	III
	Wanzhuang	0.65	0.01	15.3	19.3	III
May 2016	Shenzhuang	0.67	0.2	8.96	9.08	III
	Caizhuang	0.71	0.14	14.6	14.2	III
	Tongxinzhai	0.67	0.16	9.62	9.37	III
	Wandukou	1.08	0.14	21.9	21.4	IV
December 2016	Shenzhuang	0.86	0.38	21.1	17.9	III
	Caizhuang	0.85	0.13	0.685	0.33	I
	Tongxinzhai	0.84	0.16	9.42	8.93	III
	Wandukou	0.99	0.34	27.7	23.9	IV
	Loufan	0.84	0.37	17.1	14.9	III
May 2017	Shenzhuang	0.87	0.08	15.6	15.4	III
	Caizhuang	0.87	0.07	17.7	17.5	III
	Tongxinzhai	0.95	0.06	8.82	8.65	III
	Wandukou	0.63	0.08	8.45	8.28	III
December 2017	Shenzhuang	0.70	0.07	10.2	8.47	III
	Caizhuang	0.66	0.08	9.53	8.90	III
	Tongxinzhai	0.82	0.07	5.56	5.32	III
	Wandukou	0.66	0.07	10.80	10.40	III
	Loufan	0.74	0.07	15.60	12.80	II
May 2018	Shenzhuang	0.81	0.09	12.20	11.40	III
	Caizhuang	0.75	0.09	4.97	4.10	II

(*Continued*)

Table 8.3.   (*Continued*)

Monitoring date	Monitoring point	Permanganate index	TP	TN	NO$_3$-N	Water quality category
	Tongxinzhai	0.44	0.11	10.60	10.30	III
	Wandukou	0.81	0.09	23.30	22.70	IV
	Loufan	1.05	0.08	10.90	10.50	III
December 2018	Shenzhuang	1.25	0.08	8.99	8.83	III
December 2018	Caizhuang	1.45	0.11	4.84	4.48	II
	Tongxinzhai	1.41	0.10	4.71	4.46	II
	Wandukou	1.29	0.10	20.50	19.40	III
	Loufan	1.37	0.09	8.32	8.22	III
May 2019	Shenzhuang	0.32	0.07	3.83	3.74	II
	Caizhuang	0.44	0.07	9.49	9.28	III
	Tongxinzhai	0.36	0.09	10.10	10.00	III
	Wandukou	0.40	0.06	19.70	19.50	III
	Loufan	0.73	0.06	5.48	5.23	III
December 2019	Shenzhuang	0.32	5.53	0.02	5.47	III
	Caizhuang	0.32	6.89	0.03	6.19	III
	Tongxinzhai	0.32	10.19	0.04	9.72	III
	Wandukou	0.32	13.70	0.02	13.40	III
	Lou fan	0.48	24.22	0.02	23.50	IV

of contemporary world agricultural development, is an important support for the implementation of the concept of "public plant protection and green plant protection", and is an important measure to promote the steady growth of food production and various economic crops. It is an effective means to ensure the quantity and quality safety of agricultural products and the safety of the agricultural ecological environment. It is also an important guarantee for agricultural efficiency, farmers' income, and rural stability.

(1) Technical plan for farm household production survey

Disease, pests, and crop yield surveys: The project team held nine farmer surveys, which were conducted in the project area in September 2015, September 2016, December 2016, April 2017, July 2017, June 2018, August 2018, June 2019, and August 2019. These surveys aimed to assess the occurrence of wheat and maize diseases and insect pests during the implementation of the project (Figures 8.12 and 8.13).

In our surveys, in 2015, we randomly selected 27 wheat growers from the demonstration villages and nine wheat farmers from non-demonstration villages to evaluate the impact of the project on the occurrence of wheat diseases and insect pests. In 2016, one wheat farmer was randomly selected from the demonstration villages, and five wheat farmers from the non-demonstration villages were randomly selected to evaluate the impact

Figure 8.12. Monitoring and control of winter wheat diseases, pests, and weeds in the wheat–maize cropping system.

Figure 8.13. Monitoring and control of summer maize diseases, pests, and weeds in the wheat–maize cropping system.

of the project on wheat yield. In 2017, 2018, and 2019, eight wheat farmers were randomly selected from the demonstration villages, and 10 wheat farmers were randomly selected from the non-demonstration villages to evaluate the impacts of the project on the occurrence of wheat diseases and insect pests.

In the investigation of the diseases and insect pests in maize field, in 2015, 19 maize farmers were randomly selected from the demonstration villages in Ye County, Henan Province, and three maize farmers were randomly selected from non-demonstration villages to evaluate the impact of the project on the occurrence of maize pests and diseases. In 2017, 2018, and 2019, seven maize growers were randomly selected from the demonstration villages in Ye County, Henan Province, and one maize grower was randomly selected from non-demonstration villages to evaluate the impact of the project on the occurrence of maize diseases and insect pests.

Investigation of pest control: Nine farmer surveys were conducted in Ye County, Henan Province, in September 2015, September 2016, December 2016, April 2017, July 2017, June 2018, August 2018, June 2019, and August 2019, respectively. We monitored the use of pesticides by farmers, and the survey subjects chose the same diseases and insect pests to investigate. We mainly surveyed the number of times, types of drugs used, and the total amount of pesticides used by farmers in the demonstration villages and non-demonstration villages in 2015 and 2016 and calculated the cost of pesticides in the production costs of various crops. The survey subjects chose the same diseases and insect pest survey to evaluate the impact of project implementation on the yield of wheat and maize, while the survey subjects chose the same diseases and insect pest surveys to investigate the use of pesticides by farmers in the demonstration and non-demonstration villages. In the survey, the number of medications, the types of medications, the dosage per 667 m², and the total dosage were compared. The cost of pesticides in the production cost of various crops was calculated.

In 2015 and 2016, the large pesticide applicator and pesticide spraying drone purchased during the project implementation was not yet been applied on a large scale due to various reasons. The most common sprayer used by farmers is the conventional sprayer for spraying. In 2017, 2018, and 2019, we used the knapsack electrostatic sprayer and the ordinary knapsack electric sprayer purchased in the project implementation. However, due to other reasons such as leakage of the electrostatic sprayer in the later period, the main used tool is still the knapsack electric sprayer. Farmers in non-demonstration areas use ordinary knapsack electric sprayers.

In 2017 and 2018, a pesticide utilization test was conducted in Quanyin Village, Longquan Township, Ye County, Henan Province, used Allure Red indicator as the experimental indicator, and referring to the pesticide effective utilization rate determination process of the Institute of Plant Protection, Chinese Academy of Agricultural Sciences. The effective utilization rate of pesticides in electrostatic sprayers, large wide-width sprayers, and ordinary electric sprayers. We measured the effective utilization of pesticides in electrostatic sprayers, large-scale wide-width sprayers, and ordinary electric sprayers, respectively.

(2) Investigation results of crop diseases and insect pests
In 2015, the average time of occurrences of diseases, insects, and weeds among wheat growers in the demonstration villages was 5.778, of which 1.926, 2.074, and 1.778 were, respectively, caused by diseases, pests, and weeds. The average number of occurrences of diseases and insect pests in non-demonstration villages was 6.889 times, among which the number of diseases, insect pests, and weeds was three times, 2.333 times, and 1.556 times, respectively. The average number of occurrences of wheat diseases and insect pests in non-demonstration villages was higher than that in the demonstration villages.

In 2016, the average number of occurrences of pests and diseases in wheat growers in the demonstration villages was eight times, including three times of diseases, two times of pests and three times of weeds; the average time of occurrences of pests and diseases in non-demonstration villages was nine times, including two times of diseases, four times of pests, and three times of weeds. The average time of occurrences of wheat diseases and insect pests in non-demonstration villages is more than that in the demonstration villages. Among them, the average time of occurrences of diseases was one time less than that in the demonstration villages. The two occurrences of weeds were the same, but the average occurrence of pests in non-demonstration villages was two times more than that in the demonstration villages.

In 2017, the average number of occurrences of pests and diseases of wheat growers in the demonstration villages was five times, including two times of diseases, two times of pests, and one time of weeds; the non-demonstration villages averaged seven times pests and diseases, including three times of diseases, three times of pests, and one time of weeds. The average time of occurrences of wheat diseases and insect pests in non-demonstration villages was more than that in the demonstration villages. Among them, the average time of occurrences of diseases and insect pests was one time more than that in demonstration villages, and the occurrence of weeds was the same.

In 2018, the average number of pests and diseases of wheat growers in the demonstration villages was five times, including two times of diseases, two times of pests, and one time of weeds; the average occurrences number of pests and diseases in non-demonstration villages was 7.1 times, including 3.1 times of diseases, three times of pests, and one time of weeds. The average time of occurrence of wheat diseases and insect pests in non-demonstration villages was more than that in the demonstration villages. The average time of occurrence of diseases and insect pests was basically one time more than that in demonstration villages, and the occurrence of weeds was the same.

In 2019, the average number of occurrences of pests and diseases of wheat growers in the demonstration villages was five times, including two times of diseases, two times of pests, and one time of weeds; the average number of occurrences of pests and diseases in non-demonstration villages was 7.9 times, including 3.3 times of diseases and 3.3 times of pests. It can be seen that the average number of wheat diseases, pests, and weeds in the non-demonstration villages was more than that in demonstration villages. The average number of occurrences of diseases and insect pests was basically one to two times more than in the demonstration villages, and the average number of weeds was basically more than that in the demonstration villages. Almost once, it showed that the prevention and control of wheat diseases and insect pests in the demonstration village was better, and the demonstration effect was better.

In 2015, the average number of occurrences of pests and diseases among maize growers in the demonstration villages was five times, including two times of diseases, one time of pests, and two times of weeds; the average number of pests and diseases in non-demonstration villages was five times, including two times of diseases and one time of pests, and two times of weeds. The average frequency of maize diseases, pests, and weeds in the non-demonstration villages was the same as that in the demonstration villages.

In 2017, the average number occurrences of pests and diseases among maize growers in the demonstration villages was six times, including two times of diseases, two times of pests, and two times of weeds;

non-demonstration villages averaged six times of pests and diseases, including two times of diseases, two times of pests, and two times of weeds. The average frequency of maize diseases, pests, and weeds in non-demonstration villages was the same as that in the demonstration villages.

In 2018, the average number of occurrences of diseases, insects, and weeds among maize growers in the demonstration villages was six times, including two times of diseases, two times of pests, and two times of weeds; non-demonstration villages averaged 6.1 times of occurrences of diseases and insects, including two times of diseases, 2.1 times of pests, and two times of weeds. There was no significant difference in the average frequency of maize diseases, pests, and weeds between non-demonstration villages and demonstration villages.

In 2019, the average number of occurrences of pests and diseases among maize growers in the demonstration villages was 5.6 times, including 1.6 times of diseases, two times of pests, and two times of weeds; in non-demonstration villages, the average occurrence time of pests and diseases was 6.7 times, including two times of diseases, 2.6 times of pests, and 2.1 times of weeds. It can be seen that the average frequency of maize diseases, insects, and weeds in the non-demonstration villages was slightly higher than that of demonstration villages, indicating that the prevention and control of maize diseases and insects in the demonstration villages were better, and the demonstration effect was better.

In terms of the frequency of diseases, pests, and weeds, the frequency of maize is lower, while the frequency of wheat is higher. Generally speaking, the demonstration villages are weaker than the non-demonstration villages in terms of wheat diseases, pests, and weeds, and there is no difference between the demonstration villages and non-demonstration villages.

(3) Survey results of pesticide use
In 2015, the average application numbers of pesticides in wheat demonstration villages and non-demonstration villages were 2.8 times and 3.7 times, respectively. The average application number of pesticide applications in the non-demonstration villages was 0.9 times more than that in the demonstration villages. The average pesticide cost per 667 m^2 in demonstration villages was ¥42, and the average pesticides cost per 667 m^2 in non-demonstration villages was ¥64.2. The pesticide cost per 667 m^2 in demonstration villages was ¥22.2 lower than that in non-demonstration villages. In 2015, the frequency of occurrence of diseases, pests, and

weeds and the frequency of pesticide application in the demonstration villages were lower than those in the non-demonstration villages, but the average yield per 667 m² was 20% lower. In addition to diseases, pests, and weeds, other factors had a greater impact on yield.

In 2016, the average pesticide use in wheat cultivation in the demonstration villages and non-demonstration villages were four and 2.5 times, respectively, and the average pesticides use times in demonstration villages were 1.5 times more than that in non-demonstration villages. The pesticide cost in the demonstration villages was provided by the government, and the average pesticide cost per 667 m² in non-demonstration villages was ¥42.9. In 2016, the frequency of wheat diseases, pests, and weeds in the demonstration villages was lower than that in the non-demonstration villages, the frequency of pesticides use was 1. 5 times higher, and the yield per 667 m² was 16% higher. The project implementation effect was obvious.

In 2017, the average times of the use of pesticide use in wheat demonstration villages and non-demonstration villages were three and four times, respectively, and the average times of pesticide use in non-demonstration villages was one time more than that in the demonstration villages. The average pesticides cost per 667 m² in demonstration villages was ¥23.6, and that in non-demonstration villages was ¥34.5 per 667 m². The pesticide cost per 667 m² in the demonstration villages was ¥10.9 lower than that in the non-demonstration villages.

In 2017, the frequency of occurrence of diseases, pests, and weeds and the frequency of pesticide application in the demonstration villages were lower than those in the non-demonstration villages, but the yield per 667 m² was 10% lower. This may be caused by other factors besides diseases, pests, and weeds, such as climatic factors, insufficient application equipment, delays in the optimal application time, and uneven distribution of water and fertilizer due to different terrain heights of different plots during uniform fertilization.

In 2018 and 2019, the average times of pesticide use in wheat demonstration villages and non-demonstration villages were three and four times, respectively, and the average times of pesticide use in non-demonstration villages was one time more than that in demonstration villages. The average pesticides cost per 667 m² in the demonstration villages was ¥24, and that in non-demonstration villages was ¥36.5 per 667 m². The pesticides cost per 667 m² in the demonstration villages was ¥12.5 lower than that in the non-demonstration villages.

In 2015, the average number of times of pesticide use in maize demonstration villages and non-demonstration villages were two and three times, respectively, and the average number of times of pesticide use in non-demonstration villages was one more than that in demonstration villages. The average pesticides cost per 667 m² in demonstration villages was ¥16.3, and the average pesticide cost per 667 m² in non-demonstration villages was ¥30.99. The pesticide cost per 667 m² in the demonstration village was ¥14.68 lower than that in the non-demonstration village. In 2015, the frequency of maize diseases, pests, and weeds in the demonstration villages was the same as that in the non-demonstration villages, and the frequency of pesticide use in the demonstration villages was one time lower, and the yield per 667 m² was the same. Under the guidance of project experts, farmers in the demonstration villages reduced the amount of pesticides, which not only reduced costs, but also reduced the impact on the environment. The project implementation effect was obvious.

In 2017, the average number of pesticides in maize demonstration villages and non-demonstration villages were four and four times, respectively, and the average number of pesticides in non-demonstration villages was the same as that of the demonstration villages. The average pesticide cost per 667 m² in the demonstration villages was ¥30.0, and that in non-demonstration villages was ¥43.6 per 667 m². The pesticides cost per 667 m² in the demonstration village was ¥13.6 lower than that in the non-demonstration village. In 2017, the number of occurrences of maize diseases, pests, and weeds in the demonstration villages was the same as that in the non-demonstration villages, and the frequency of medication use in the demonstration villages was the same as that in the non-demonstration villages. It was ¥13.6 lower than the non-demonstration villages. In 2017, the number of occurrences of maize diseases, pests, and weeds in the demonstration villages was the same as that of the non-demonstration villages, and the frequency of medication use in the demonstration villages was the same as that in the non-demonstration villages. It was ¥13.6 lower than the non-demonstration villages.

In 2018 and 2019, the average number of pesticides in maize demonstration villages and non-demonstration villages were four and four times, respectively, and the average number of pesticides in non-demonstration villages was the same as that in the demonstration villages. The average pesticide cost per 667 m² in the demonstration villages was ¥31, and that in non-demonstration villages was ¥44 per 667 m². The pesticide cost per

667 m² in the demonstration village was ¥13 lower than that in the non-demonstration village. This was because the project team has promoted electrostatic sprayers in the demonstration area and replaced some traditional electric sprayers, making the pesticide utilization rate in the project area higher than that in the non-project area, thereby reducing the amount of pesticide use per unit area and reducing the cost of pesticide use in the project area. Under the guidance of project experts, farmers in the demonstration villages reduced the number of pesticides, which not only reduced costs but also reduced the impact on the environment. The project implementation effect was obvious.

In general, the application frequency of pesticides for wheat and maize in the demonstration villages is lower than that in non-demonstration villages, and the pesticide cost is also lower than in the non-demonstration villages.

(4) Investigation report on the use of medical devices
In 2015, both the wheat demonstration villages and non-demonstration villages used conventional sprayers for pesticide application. The occurrence time of diseases, pests, and weeds and the use of pesticides in the demonstration villages were lower than those of the non-demonstration villages, but the yield per 667 m² was 20% lower. In addition to diseases, pests, and weeds, other factors had a greater impact on yield.

In 2016, both the wheat demonstration villages and non-demonstration villages used conventional sprayers for pesticide application. The occurrence time of wheat diseases, pests, and weeds in the demonstration villages was one time lower than that in the non-demonstration villages, the application time was 1.5 times higher, and the yield per 667 m² was 16% higher. The project implementation effect was significant.

In 2017, 2018, and 2019, wheat cultivation in the demonstration area of Ye County, Henan Province, used the knapsack electrostatic sprayer, and the wheat farmers in the non-demonstration area used the ordinary knapsack electric sprayer. The test results showed that the effective utilization rate of pesticides using knapsack electrostatic sprayer (62.7%) was higher than that of ordinary knapsack electric sprayer (52.4%) in non-demonstration areas. It showed that to achieve the same control effect, the average pesticide use times in non-demonstration areas should be one more time than in demonstration areas, which made the average pesticide cost per 667 m² (¥34.5 per 667 m²) in non-demonstration areas also higher than the cost of medication in demonstration villages (¥23.6 per 667 m²).

In 2015, both maize demonstration villages and non-demonstration villages used conventional sprayers for pesticide application. The frequency of maize diseases, pests, and weeds in the demonstration villages was the same as that in the non-demonstration villages, and the frequency of pesticide application in the demonstration villages was one time lower, and the average yield per 667 m² was the same. Under the guidance of project experts, farmers in the demonstration villages reduced the amount of pesticides, which not only reduced costs but also reduced the impact on the environment. The project implementation effect was obvious.

In 2017, 2018, and 2019, the spraying method used for maize in the Ye County Demonstration Zone in Henan Province was a knapsack electrostatic sprayer, while the spraying method used in the non-demonstration area was an ordinary knapsack electric sprayer. The test results showed that the effective utilization rate of pesticides using knapsack electrostatic sprayers (62.7%) was higher than that of ordinary knapsack electric sprayers in non-demonstration areas (52.4%), thus reducing the amount of pesticides used per unit area and therefore the pesticide use cost in the project area. Under the guidance of project experts, farmers in the demonstration villages reduced the amount of pesticides, which not only reduced costs, but also reduced the impact on the environment. The project implementation effect was obvious.

The effective utilization rate of pesticides using the large wide-width pesticide applicator was higher than that of the ordinary knapsack electric sprayer. The effective utilization rate of pesticides using the knapsack electrostatic sprayer was higher than that using the ordinary knapsack electric sprayer. The effective utilization rate of pesticides of the knapsack electrostatic sprayer was higher than that of the large wide-width pesticide applicator (Figure 8.14). The use of high-efficiency pesticide equipment can reduce the amount of pesticides used per unit area and reduce the pesticide cost.

(5) Crop yield impact assessment report
In 2015, the average yield of wheat in the demonstration villages was 378 kg per 667 m², the average yield of non-demonstration villages was 478 kg per 667 m², and the average yield of demonstration villages was 99 kg per 667 m² lower than that of non-demonstration villages (Figure 8.15). In 2016, the average yield of wheat in the demonstration village was 410 kg per 667 m², and the average output in the non-demonstration village was 355 kg per 667 m². Compared with non-demonstration villages,

Figure 8.14.   Pesticide machinery in the wheat–maize cropping system.

Figure 8.15.   Winter wheat production measurement in the wheat–maize cropping system.

the average yield of the demonstration village was 55 kg per 667 m² higher than that in the non-demonstration village. In 2017, the average yield of wheat in the demonstration villages was 300 kg per 667 m², and the average yield of non-demonstration villages was 325 kg per 667 m². The average yield of demonstration villages was 25 kg per 667 m² lower than that of the non-demonstration villages. In 2018, the average yield per 667 m² of wheat in the demonstration village was 330 kg, and the average yield per 667 m² in the non-demonstration village was 310 kg. The average yield per 667 m² in the demonstration village was 10 kg higher than that in the non-demonstration village. In 2019, the average wheat yield per 667 m² in the demonstration villages was 345 kg, and the average yield per 667 m² in the non-demonstration villages was 308 kg. The average yield per 667 m² in the demonstration villages was 37 kg higher than that in the non-demonstration villages. In 2015, the average maize yield of the demonstration villages and non-demonstration villages was 550 kg per 667 m². In 2017, the average yield of maize in both the demonstration and non-demonstration villages was 580 kg per 667 m². In 2018, the average maize yield per 667 m² in the demonstration and non-demonstration villages was 590 kg. In 2019, the average maize yield per 667 m² in the demonstration villages was 610 kg, and the average maize yield per 667 m² in non-demonstration villages was 596 kg. According to investigation and analysis, the project area produced different degrees of impact on crop yields through the integration and demonstration of key technologies for crop production, emission reduction, and carbon increase, innovation and application of supporting policies, and expansion and improvement of public knowledge. Compared with the non-project area, the overall output of the project area was higher than that of the non-project area.

In 2015, the average wheat yield in the demonstration villages was lower than that in the non-demonstration villages. Based on the occurrence of diseases and insect pests, this was mainly due to the higher incidence of diseases and insect pests in the non-demonstration villages than in the demonstration villages. In 2016, the yield of wheat in the demonstration villages was higher than that in the non-demonstration villages. The average wheat yield in the demonstration villages in 2017 was slightly lower than that in the non-demonstration villages. This may be caused by other factors besides diseases, insects, and weeds, such as climatic factors, insufficient application equipment in the demonstration area, delays in the optimal application time, and different areas in the demonstration area. The different terrain heights of the blocks lead to uneven distribution of water and fertilizer. In 2018 and 2019, the average yield of wheat in the demonstration villages was higher than that in the non-demonstration villages, indicating that the demonstration technology has been better promoted in the project area.

In 2015, the yield per 667 m^2 of maize in the demonstration villages was the same as that in the non-demonstration villages, and the number of pests and diseases in the demonstration villages affected the yield more than in the non-demonstration villages. In 2017, the yield per 667 m^2 of maize in the demonstration villages was the same as that in the non-demonstration villages, but pesticide cost per 667 m^2 was lower. The pesticides cost per 667 m^2 in the demonstration villages was ¥13.6 lower than that in the non-demonstration villages. This is because the project team promoted electrostatic sprayers in the demonstration area to replace some traditional electric sprayers, making the pesticide utilization rate in the project area higher than that in the non-project area, thereby reducing the amount of pesticide use per unit area and reducing the cost of pesticide use in the project area. In 2018, the yield per 667 m^2 of maize in the demonstration villages was the same as that in the non-demonstration villages, but the pesticide cost per 667 m^2 was lower. The pesticides cost per 667 m^2 in the demonstration villages was ¥13 lower than that in the non-demonstration villages. In 2019, the yield per 667 m^2 of maize in the demonstration villages was slightly higher than that in the non-demonstration villages, and the pesticides cost per 667 m^2 was lower. This is because the project team promoted electrostatic sprayers in the demonstration area to replace some traditional electric sprayers, making the pesticide utilization rate in the project area higher than that in the non-project area, thereby reducing the amount of pesticide use per unit area and reducing the cost of pesticide use in the project area.

The implementation of the project has affected crop yields to varying degrees, mainly affecting the average yield per 667 m^2 of different crops in the demonstration villages and non-demonstration villages, and there was no stable relationship between crop yields between the demonstration villages and non-demonstration villages. The two expected yield outcomes for individual farmers in the project area are contrary to the actual situation. The possible reasons for this situation are: natural factors, such as floods, droughts, sudden weather effects; human factors, such as farmers' acceptance of new technologies, the enthusiasm not being high, and they continuing to plant according to the original planting technology; lack of comprehensive quality of farmers, and the incomplete understanding of national policies or new technologies make the technology not promoted in accordance with the expected thinking. In view of these possible factors, we should proceed from both objective and subjective aspects, and further promote new technologies to increase crop yields and achieve sustainable agricultural development.

(6) Questions and suggestions

(i) The impact of non-project factors on project implementation effects and suggestions. The impact of project crop yields may be affected by farmers' planting habits, floods, droughts, various meteorological disasters, and human factors. The evaluation of project implementation effects requires a comprehensive consideration of various factors. As for a comprehensive impact, it is suggested that other factors affecting crop yield should be taken into consideration, and the impact of non-project factors on the implementation effect of the project should be stripped off so as to scientifically evaluate the implementation effect of the project.

(ii) Recommendations on project implementation. The survey found that the pesticides use by farmers in the demonstration and non-demonstration villages are more complicated. In future project implementations, it is necessary to scientifically evaluate the rationality of pesticide use by farmers in demonstration and non-demonstration villages based on actual conditions.

(iii) Selection of survey objects. In the project implementation, due to the number of households surveyed, the cleanliness between the demonstration villages and the non-demonstration villages between the years are poor, and the number of households and planting scales are different. It is recommended to select fixed farmers for long-term monitoring to improve the scientific nature of project research.

(iv) The use of medical equipment. The large-scale wide-width medical applicator and drones newly purchased in the project area should be used on a large scale as soon as possible to conduct the investigation of the efficacy and use of the new equipment in the project area.

8.3.4 *Social Impact Monitoring and Evaluation*

(1) Project background and research situation

As the international community pays more and more attention to climate change, greenhouse gas emission reduction, and food security, research on farmland soil carbon sequestration and emission reduction technology has received unprecedented attention from the scientific community and these have gradually been widely used in agricultural production in various countries. Climate-smart agriculture aims to improve the nitrogen fixation capacity of farmland soil without reducing the level of crop productivity, while reducing field greenhouse gas emissions, and adopting diverse cultivation management modes to enhance crop production's adaptability to climate change.

China's climatic conditions, land resources, and planting systems all have obvious regional characteristics. Carbon sequestration and emission reduction technologies have different requirements and effects in various regions, and certain management measures are difficult to continue to promote due to their impact on production. Wheat, rice, and maize are the three main grain crops in my country, and their total yield accounts for more than 85% of China's grain yield. The main grain production areas in North China and East China are responsible for ensuring food security. The sown area and grain yield of grain crops account for 63% and 67% of the country's total grain crop area and yield, respectively. At the same time, major food production areas are also facing serious organic carbon losses, large amounts of nitrogen fertilizers and other problems, as well as an urgent need for carbon sequestration and huge potential for energy-saving and emission reduction of greenhouse gases. Therefore, the promotion and application of energy-saving and carbon sequestration technologies under the premise of ensuring food production in major grain production areas, as well as demonstration and evaluation of emission reduction effects, can not only improve soil fertility and productivity, and slow down the emission of greenhouse gases in the soil, but also maintain agriculture in my country. Sustainable development is a strategic choice.

Ye County is located in Pingdingshan City, southwest of central Henan Province. It is located at the junction of Huanghuai Plain and Funiu Mountain. The county has a total area of 1,387 square kilometers, jurisdiction over 18 towns, and a population of 868,000. The project area, Longquan Township, is located in the southeast of Ye County. The project areas are all plain areas, and the project villages are basically arranged on both sides of the Yanli River. The annual average temperature is 14.9°C, the highest month (July) average temperature is 27.5°C, the lowest month (January) is 1.3°C; the annual precipitation is 778.9 mm, the highest precipitation month (July) is 177.7 mm, and the lowest precipitation month

(January) is 15.7 mm. The frost-free period averages 217 days. There are three types of soil: yellow brown loam soil, sand ginger black soil, and fluvo-aquic soil.

In 2017, we visited five project villages in Ye County, Henan Province (Figure 8.16). From each village 20 households were selected out of five villages in Longquan Township: Loufan Village, Niuduzhuang Village, Guoluzhuang Village, Beidaying Village, and Muzhuang Village. In order to conduct a comparative analysis of the project effects, four non-project villages in Yeyi Town were selected: Caizhuang Village, Shenwan Village, Tongxinzhai Village, Wandukou Village, and Longquan Township, and six non-project villages: Nandaying Village, Baihaozhuang Village, Shenzhuang Village, Quanyin Village, Longquan Village, and Zhongzhang Village. From each non-project village, 10 households were selected. The project team members communicated with village leaders and filled out village-level questionnaires, and learned in detail about the socioeconomic conditions, crop planting conditions, land transfer conditions, and cooperative development conditions of five project villages and 10 non-project villages and received a total of 200 valid questionnaires. In 2018, 14 villages in Longquan Township and Yeyi Town were selected in Yexian County, including Loufan Village in Longquan Township, Niu Muzhuang Village, Beidaying Village, Guoluzhuang Village and Ximu Village, and Shenwan Village in Yeyi Town. A total of 190 valid questionnaires were received from 12 project villages in Wandukou and other villages, and two non-project villages in Banjielou and Xinshanzhuang Village in Longquan Township. In 2019, eight project team members went to Ye County, Henan Province, to investigate the impact of climate-smart major food crop production projects on local socioeconomic development and ecological development. This survey focused on eight villages in Longquan Township in Ye County, including Nandaying Village, Xiaohewang Village, Banjielou Village, and Xinshanzhuang Village. A total of 112 valid questionnaires were collected.

(2) Evaluate the economic and social effects of project implementation
(i) Stabilize production and reduce input, and promote increased production and income

The yield of major food crops rose steadily. The overall wheat yield in Ye County from 2016 to 2019 showed an overall upward trend. The 4-year average yield in the project area was 6,304 kg hm^{-2}, and it

Figure 8.16.　Survey of farmers in the wheat–maize cropping system.

increased at a rate of 652 kg hm^{-2} per year; the 4-year average wheat yield in the non-project area was 6,004 kg hm^{-2}, and the growth rate is 462 kg hm^{-2} per year; the difference in wheat yield between the project area and the non-project area in 2016–2019 is 190 kg hm^{-2} per year on average. The overall maize in Ye County declined first and then the average yield in 2016–2019 in the project area and the non-project area is 6,521 and 6,229 kg hm^{-2}, respectively. The yield difference in maize between the project area and the non-project area in 2016–2019 is on the rise, and increasing at a rate of about 111 kg hm^{-2} per year. From 2016 to 2019, the annual wheat–maize yields in the Ye County project area and non-project areas both showed an upward trend, with an average annual increase of 717 and 416 kg hm^{-2}, respectively, and the 4-year average annual yields were 12,825 and 12,233 kg hm^{-2}, respectively. The project area is 592 kg hm^{-2} higher than the non-project area on average (Table 8.4).

The planting costs of major food crops have gradually decreased. From 2016 to 2018, the wheat planting cost in the project area and non-project area of Ye County decreased by an average of ¥216 and ¥209 ha^{-1} yr^{-1}, respectively, and the average cost of wheat in the project area was

Table 8.4.   2016–2019 Ye County project area and non-project area main crop yield per unit area (kg hm^{-2}).

Year	Wheat			Maize			Wheat–Maize		
	Project	Non-project	D-value	Project	Non-project	D-value	Project	Non-project	D-value
2016	5,190	5,123	68	6,480	6,450	30	11,670	11,573	98
2017	6,368	6,127	241	6,509	6,136	373	12,877	12,264	614
2018	6,259	6,091	168	6,344	5,985	359	12,603	12,076	527
2019	7,400	6,675	725	6,750	6,345	405	14,150	13,020	1,130

Table 8.5.   The total planting cost of main crops in the project area and non-project area of Ye County from 2016 to 2019 (¥ ha^{-1}).

Year	Wheat			Maize			Wheat–Maize		
	Project	Non-project	D-value	Project	Non-project	D-value	Project	Non-project	D-value
2016	5,400	5,400	0	4,800	4,845	−45	10,200	10,245	−45
2017	5,401	5,500	−99	4,858	4,868	−9	10,259	10,367	−108
2018	4,969	4,982	−13	4,848	4,931	−83	9,816	9,913	−97
2019	5,343	5,024	319	4,793	4,887	−95	10,135	9,911	225

¥5,260 and ¥5,290 ha^{-1}, respectively. The overall wheat planting cost in Ye County has suddenly risen, resulting in a 7.5% and 0.8% increase in costs in the project area and non-project area, respectively, compared to 2018. From 2016 to 2019, the overall maize planting cost in Ye County project area and non-project area did not change much, with the 4-year average cost being ¥4,825 and ¥4,883 ha^{-1}, respectively. The planting cost in the project area was relatively low. From 2016 to 2019, the annual wheat–maize rotation system planting cost in the Ye County project area and non-project area decreased by an average of ¥64 and ¥146 ha^{-1}, respectively, and the 4-year average cost was ¥10,103 and ¥10,109 ha^{-1}, respectively. The planting cost in the project area is relatively low (Table 8.5).

The overall decline in the production cost of major food crops reflects the positive results of the climate-smart food project, however, the agricultural production cost is still at a relatively high level. The higher agricultural production input in the project area ensures food production, on the one hand, and it also increases the expenditure burden of farmers, on the other hand, and limits the continuous increase in agricultural operating income.

The income per unit area of major food crops is stable and improving. From 2016 to 2019, the income per unit area of wheat and maize in Ye County showed an overall upward trend. The growth rate of wheat yield per unit area in Ye County project area and non-project area was ¥1,290 and ¥850 ha^{-1} respectively, and the 4-year average yield per unit area was ¥8,135 and ¥7,450 ha^{-1}, respectively. The income difference between the project area and the non-project area is also on the rise, with an average growth rate of ¥440 ha^{-1} yr^{-1}. The yield of maize per unit area in Ye County project area and non-project area fluctuated. The 4-year average yield per unit area was ¥5,941 and ¥5,365 ha^{-1}, respectively; however, the difference in yield per unit area of maize showed an upward trend, with an average annual increase of ¥160 ha^{-1}. The Ye County wheat–maize rotation system project area grew at an average annual rate of ¥1,545 ha^{-1}. The non-project area changed little, and only saw a large increase in 2019, an increase of 22.7% compared with 2018. The income difference between the project area and the non-project area has increased at an average annual rate of ¥600 ha^{-1}. In 2019, it has reached ¥2,440 ha^{-1}. The 4-year average income of the project and non-project areas is ¥14,076 and ¥12,815 ha^{-1} (Table 8.6).

The degree of agricultural mechanization has gradually increased and production efficiency has greatly improved. In recent years, with the support of the project and national agricultural machinery subsidies,

Table 8.6.   2016–2019 Ye County project area and non-project area main crops unit area income (¥ ha^{-1}).

	Wheat			Maize			Wheat–Maize		
Year	Project	Non-project	D-value	Project	Non-project	D-value	Project	Non-project	D-value
2016	6,750	6,615	135	5,865	5,610	255	12,615	12,225	390
2017	7,208	6,632	576	5,556	4,950	606	12,765	11,583	1,182
2018	7,674	7,322	352	5,684	5,004	680	13,357	12,326	1,031
2019	10,910	9,231	1,679	6,660	5,895	765	17,570	15,126	2,444

the mechanization rate of the project county has continued to increase. The total power of agricultural machinery in Ye County has reached 630,000 kilowatts and the machine harvest area has reached 99,000 hectares. With the continuous increase in agricultural machinery, agricultural socialized services in the form of agricultural machinery cooperatives and individuals have flourished.

According to the research situation, machine farming, machine seeding, and machine yield rate of wheat in Ye County is above 85%, and machine farming, machine seeding, and machine yield rate of maize is above 80%, and the overall mechanization rate is higher. Especially in the project villages where the overall circulation is carried out, the production of the machinery assembly has reached almost 100%, which provides a solid guarantee for high-efficiency and high-quality food production. The mechanization of agricultural production and social services have improved the efficiency of agricultural production, while greatly reducing the labor burden of farmers, saving labor time, and increasing farmers' social welfare.

(ii) Reduce fertilizers and pesticides to promote energy conservation and emission reduction

The use of chemical fertilizers for major food crops is decreasing year on year. From 2016 to 2018, the fertilizer consumption of wheat in Ye County showed an overall downward trend, but it increased in 2019. The fertilizer consumption of wheat in the project area and non-project area of Ye County decreased at a rate of 18 and 30 kg ha^{-1} yr^{-1}, respectively. The average fertilizer consumption in 3 years is 800 and 810 kg ha^{-1}, respectively. From the difference in the consumption, it can be seen that the wheat fertilizer consumption in the project area from 2016 to 2018 is slightly lower than that in the non-project area. From 2016 to 2019, the fertilizer consumption of maize in Ye County showed a downward trend, especially during 2016–2017. The downward trend was particularly obvious. From 2016 to 2019, the project area and non-project area decreased at a rate of 116 and 95 kg ha^{-1} yr^{-1}, respectively. The average fertilizer consumption was 933 and 918 kg ha^{-1}, respectively; the gap between the project area and the non-project area in the use of maize fertilizer in 2016–2019 gradually reduced, and the decline was significant in 2016–2017, although the consumption in the project area was lower in 2017–2019. But the decline is not obvious. From 2016 to 2019, the fertilizer consumption of wheat–maize rotation system in Ye County showed a downward trend. The project area and non-project area decreased at a rate of 113 and 106 kg ha^{-1} yr^{-1}, respectively, and the decline rate was faster in the project area; this is the same as the maize fertilizer consumption situation. The difference in fertilizer use between the project area and the non-project area in the wheat–maize annual crop rotation system declined significantly from 2016 to 2017, and it slowed down from 2017 to 2018. In 2018–2019, the use of fertilizer in the project area was higher than that in the non-project area (Table 8.7).

On the whole, the reduction in the use of chemical fertilizers reflects the key part of the project — the successful demonstration and application of the farmland chemical fertilizer reduction application technology. However, in absolute terms, the use of chemical fertilizers in the project area is still relatively large, and further reductions are needed.

The use of pesticides in major food crops is decreasing year on year. From 2016 to 2019, the frequency of wheat pesticide use in Ye County showed a general downward trend. From 2016 to 2017, the decline was large. The project area and non-project area decreased by 0.5 and 1.7 times, respectively; the number of pesticide applications in the project area did not change much from 2017 to 2019. The number of medications in the project area is on the rise, increasing at a rate of 0.2 times per year; the average number of medications in the project area and non-project area in 2017–2019 is 2.5 times, and the gap in the number of medications gradually decreases at 0.2 times per year. From 2016 to 2019, the Ye County maize project area and non-project area both decreased at a rate of 0.1 times per year. The average 4-year number of medications was 2.2 times. The overall number of medications in the non-project area was relatively small. But between the project area and the non-project area, the gap in the number

Table 8.7.   The fertilizer consumption per unit area of main crops in Ye County project area and non-project area from 2016 to 2019 (kg ha^{-1}).

	Wheat			Maize			Wheat–Maize		
Year	Project	Non-project	D-value	Project	Non-project	D-value	Project	Non-project	D-value
2016	825	848	−23	1,215	11,48	68	2,040	1,995	45
2017	790	790	0	845	848	−3	1,635	1,637	−2
2018	788	790	−1	841	842	−2	1,629	1,632	−3
2019	836	811	24	830	834	−4	1,666	1,645	21

Table 8.8.   The number of main crops used in the project area and non-project area of Ye County from 2016 to 2019 (times).

	Wheat			Maize			Wheat–Maize		
Year	Project	Non-project	D-value	Project	Non-project	D-value	Project	Non-project	D-value
2016	3.0	4.0	−1.0	2.4	2.3	0.1	5.4	6.3	−0.9
2017	2.5	2.4	0.1	2.2	2.1	0.1	4.7	4.5	0.2
2018	2.4	2.4	0.1	2.2	2.1	0.1	4.6	4.5	0.1
2019	2.5	2.7	−0.2	2.0	2.1	−0.1	4.5	4.8	−0.3

of medications is decreasing at a rate of 0.05 times a year. From 2016 to 2019, the frequency of medication use in the wheat–maize rotation system project area and non-project area of Ye County decreased at an average rate of 0.3 and 0.5 times per year, respectively. The average frequency of medication use in the 4 years was 4.8 and 5.0 times (Table 8.8).

After the implementation of the climate-smart major food crops project in Ye County from 2016 to 2019, the fertilizer consumption of the wheat–maize rotation system after the implementation of the project was 2. 5 kg ha⁻¹, lower than that in the non-project area. The difference in fertilizer consumption shows that the gap between the two is still gradually getting bigger. At the same time, the number of pesticides used in the system has continued to decrease after the implementation of the project. Although the current consumption is still higher than that in non-project areas, the trend of pesticides use in non-project areas is not stable enough. It dropped sharply in 2016–2017 and rebounded in 2017–2018.

The rate of returning straw to the field has increased, and farmers have increased their willingness for green and sustainable development. Returning straw to the field is a key technology of the climate-smart main food crop project. It is not only a measure to increase production of fertility, but also a core technology for carbon sequestration and emission reduction, and the effect of emission reduction is obvious. Straw contains a lot of fresh organic materials, which can be converted into organic matter and quick-acting nutrients after a period of decomposing after being returned to the farmland. Returning straw to the field can increase soil organic matter, improve soil structure, loosen the soil, increase porosity, reduce capacity, promote microbial activity and the development of crop roots, improve soil's physical and chemical proper-ties, and supply potassium and other nutrients. Returning straw to the field can promote agricultural water saving, cost saving, increase in production and efficiency. Returning straw to the field has a significant effect on increasing fertilizer and yield, generally increasing yield by 5–10%. Since the implementation of the project, the project area has achieved a 90% return rate of straw, which is very effective.

The farmers' concepts of science, technology, and environmental pro-tection have gradually increased, and their willingness to save energy and reduce emissions has continued to increase. In the survey in Huaiyuan County, 81% of the farmers indicated that they would accept a unified fertilization formula; 80% of the farmers were willing to accept a unified pest control, and 75% of the farmers indicated that climate change has an impact on agricultural production, but most of them are negative; 83% of farmers believe that agricultural technology is very important for solving current agricultural problems. In the Ye County survey, 78% of the farm-ers expressed their willingness to accept a unified fertilization formula; 83% of the farmers expressed their willingness to accept a unified pest control, and more than 60% of the farmers expressed the need to lose weight and reduce medicines in agricultural production; 90% of the farm-ers believed in new agricultural technology. It is very important to solve agricultural production. On the whole, farmers in the project area have greatly improved their awareness of agricultural science and technology and farmers' awareness of environmental protection. They have accepted the concept of climate-smart food crop projects from a cognitive perspec-tive and are willing to follow in action.

(iii) Drive disadvantaged groups to help fight poverty

The climate-smart main food project actively combines the national rural revitalization strategy and targeted poverty alleviation work and attaches great importance to the care of vulnerable groups; during the implementa-tion of the project, the maximum amount of assistance is provided to vulnerable groups such as women, the elderly, the disabled, and poor households. Among the farmers participating in the climate-smart food crop project in Ye County, the poor accounted for 8.5%, and the poor accounted for 25% of the farmers participating in the project training, and accounted for about 30.5% of the trained personnel. Amony those who participated in the training, the proportion of women is as high as 54%; the average number of employment in the village is 18 people with disabilities, and their per capita income can reach more than ¥4,800. The average number of women employed in the village is 240, and the per capita income can reach ¥5,300. The average employment of the elderly is 130 people and the per capita income can reach ¥4,140. It has played an important role in increasing the income of disadvantaged groups and improving their social welfare. It is more helpful for poverty alleviation and has elicited a good social response.

Chapter 9

Summary of China's Climate-Smart Agriculture Mode Experience

Abstract

China's agriculture is in an important transition period. In order to make a long-term impact, we emphasize on energy conservation and emission reduction, adapting to and mitigate climate change, and exploring the development path of climate-smart agriculture in China to be of great practical significance for guaranteeing China's food security, realizing farmers' poverty alleviation, and achieving the goals of implementing the climate change agreement. China's climate-smart major food crop production projects have explored innovative institutional approaches such as establishing subsidy policies, technology demonstration, publicity and training, and large-scale production for carbon sequestration and climate change adaptation, laying a good foundation for the development of climate-smart agriculture. It has been proved that climate-smart agriculture is an important mode of green agricultural development in China and can be regarded as a sustainable agricultural development mode integrated into the internationalization. The core of promoting climate-smart agriculture into the mainstream mode of Chinese agriculture is to further promote agricultural subsidy policy and the innovation of technology mode. Based on the experience of foreign countries, the establishment and improvement of China's climate-smart agriculture policy system, technology, theoretical system, and popularization mechanism are important measures to promote China's climate-smart agriculture development. Specifically, China's climate-smart agriculture needs support from the following aspects: (1) climate-smart agriculture development planning and supporting incentive policies and measures; (2) national stable and sustainable financial support for climate-smart agriculture technology research and development and model integration; (3) strengthening the integration, demonstration, and application of climate-smart agriculture production technology; (4) developing climate-smart agriculture model according to local conditions; and (5) strengthening international cooperation and communication of climate-smart agriculture.

9.1 Challenges to the Development of CSA in China

China's climate-smart agriculture (CSA) is still in its infancy, and the relevant policy system, system construction, and incentive mechanism need to be further improved. The relevant policies of CSA in China are far from the climate change agriculture, and a series of policies have been introduced respectively in the fields of agricultural production to cope with climate change and green and sustainable agricultural development to promote the implementation of policies in related fields. For example, the *National Strategy for Adaptation to Climate Change* issued in 2013 had made it clear that agricultural adaptation to climate change is the direction of efforts, including strengthening agricultural monitoring and early warning, disaster prevention and mitigation measures, enhancing the adaptability of crop farming, and strengthening the security for agricultural development. *The Work Plan for Controlling Greenhouse Gas Emissions during the 13th Five-Year Plan Period* (2016–2020), issued in 2016, made it clear that China's agricultural development should be enhanced at the same time as the ecosystem function of the system. In addition, CSA is a comprehensive science, the related theory and technical support system has not been improved, and the related research needs to be further strengthened.

At present, the main factors affecting the application of CSA are the characteristics of population structure and development status of rural society in China. China's agricultural production is still dominated by small farmers. In the past 20 years, China's urbanization has developed rapidly, resulting in a large outflow of young and middle-aged rural population. Women and the elderly have become the main labor of agricultural production. The research results of the project showed that due to the limitation of various reasons such as low educational level, rural groups, mainly women and the elderly, paid less attention to new agricultural concepts and new technologies, and had a lag in understanding and acceptance of CSA. In addition, the proportion of agricultural income in the total household income has been decreasing, and farmers' expectation of the income from farming has been declining, which has also affected the popularization and application of climate-smart agriculture to some extent.

9.2 Inspiration of CSA to China's Agricultural Development

At present, China's agriculture is in an important transition period, which is also an opportunity period for sustainable development of agriculture. Against the background of climate change assuming importance in the world, how to make a long-term layout, emphasize energy conservation and emission reduction, adapt to and mitigate climate change, and explore the path of CSA development in China is of great practical significance for ensuring China's food security, lifting farmers out of poverty, and achieving the goals of implementing climate change.

9.2.1 *Establishment and Improvement of China's CSA Policy System*

To sum up, national CSA policies include the following four factors. The first is the specific background of the baseline condition assessment, including assessment of the effects of climate change on agriculture domain-specific functions and their vulnerability, the existing agricultural system and implementation of an intelligent agricultural climate, costs and benefits, the implementation of an intelligent agricultural climate barriers, and the current climate action on the implementation of an intelligent agriculture. The second is the strong multi-stakeholder institutions. Establish a fixed platform for political decision makers, domestic participants, and international partners to analyze relevant risks and discuss policy priorities, so as to ensure the effective implementation of climate-smart agriculture. Among them, transparent and credible information analysis is one of the important roles of multi-stakeholder institutions. The third is the coordination framework. Adaptation, climate change mitigation, and food security require the participation of governments, producers, businesses, and international partners. The central government can establish a policy framework to coordinate the public and private sectors, and effectively implement climate-smart agriculture, such as market incentives, financing mechanisms, and technical assistance. At the same time, the framework should clearly coordinate the relationship between agriculture, forestry, and land use in order to avoid conflicts between different sectors. The fourth is a multi-scale information system. Climate-smart agriculture increases agricultural productivity and the achievement of GHG reduction targets requires a high degree of risk assessment, vulnerability assessment, and context-specific strategies. To achieve climate-smart agricultural benefits, government needs information at multiple scales, including research and development, advisory services, information technology, and monitoring and assessment. The output of improved technologies and methods for climate-smart agriculture-related projects needs to be recorded and reported for the purpose of motivating subsequent participants and testing the technologies and methods. Information systems also require the costs and benefits of climate-smart agriculture in all its aspects in order to facilitate the establishment and improvement of relevant policies.

9.2.2 *Establishment of China's CSA Technology, Theoretical System, and Market Promotion Mechanism*

9.2.2.1 *Establish China's CSA Technology and Theoretical System*

Although the concept of CSA has not been put forward for a long time, many countries have already tried and applied specific emission reduction and carbon sequestration models to cope with climate change, such as Australia's efficient emission reduction model, the United States' fallow carbon sequestration model, the crop rotation model, and the EU's systemic response model. At present, China also has many low-carbon agricultural development models, such as ecological agriculture, organic agriculture, and green agriculture. According to the climate type and agricultural structure, these agricultural development models are sorted into specific technological models for different regions to cope with climate change, thus comprehensively forming China's CSA technology system. Taking the mode of rice cultivation as an example, in the main rice growing areas of China, a cultivation technology system has been initially formed, which is based on the key technologies of full return of straw mulch, rotary soil tillage, mechanized dry direct seeding, deep fertilization with fertilizer orientation, infiltration irrigation with box ditch, and comprehensive control of diseases, pests, and weeds. This new mode of rice cultivation integrating "conservation tillage, infiltration irrigation, and precise fertilization" has significantly reduced the greenhouse gas emissions of rice fields. At the same time, according to the corresponding technology model, the monitoring methodology, standards, and the theoretical system of China's CSA is constructed.

9.2.2.2 Establish the CSA Demonstration Zone

We should draw on the advanced experience of CSA at home and abroad, integrate the technological models related to high-yield and emission reduction, including the measures to cope with ordinary and extreme climates, establish CSA pilot projects adapted to different climatic regional types, and demonstrate and popularize them in relevant regions. At the national and regional scale, the allocation, development direction, and development mode of agriculture should be determined scientifically according to the natural and economic conditions of different regions, including soil and water resources, natural and climatic conditions, agricultural production and operation modes, and economic development level. The establishment of CSA system includes the establishment of agro-meteorological disaster monitoring and early warning and control service system, the improvement of meteorological disaster prevention and reduction early warning system, the prediction and prevention of meteorological conditions for the occurrence of agricultural diseases and insect pests, the integration of climate-smart technology and mode, and the development of agro-meteorological disaster insurance system. Depending on large specialized farmers of planting and breeding, specialized cooperatives and other new types of agricultural operators, we should develop agricultural operations on an appropriate scale, establish pilot demonstration zones for CSA, and select and breed high-quality seeds with high yield, efficiently use nutrients, and promote strong resistance to stress according to local conditions. Besides, it is worthy to breed low emission and high-quality breeds of livestock and poultry; use water-saving measures, engage in pesticide and fertilizer reduction, and efficiency technology, comprehensive utilization of farmland waste (straw, agricultural film, livestock and poultry dung); and develop the characteristic technology model and industry of each region according to local conditions.

9.2.2.3 *Establish Marketing Approaches Based on Green Value Chains and the Private Sector*

Depending on the government, improve climate-smart agricultural infrastructure and equipment in the early stage, and establish climate-smart organization framework and cooperation mechanism. Relying on the market mechanism, improve the interest connection mechanism, promote the development of climate-smart agriculture-related industries, and achieve climate-smart agriculture with independent participation and sustainability. For example, cultivate the carbon trading market mechanism of climate-smart agriculture and explore the agriculture-smart investment and public–private partnership (PPP) model. Agricultural carbon trading is still in its infancy and is a weak link in the field of international carbon trading. At present, Brazil, Mexico, the Philippines, and India have a large number of agricultural carbon trading projects. Therefore, we can jointly carry out targeted cooperation on agricultural carbon trading projects to sum up experience and lessons, so as to provide useful reference for market development mechanisms of climate-smart agriculture.

9.3 Prospect of CSA in China

China's climate-smart major food crop production projects have explored innovative institutional approaches such as establishing subsidy policies for carbon sequestration and climate change adaptation, technology demonstration, publicity and training, and large-scale production, laying a good foundation for the development of climate-smart agriculture. This project has completed the *Climate-Smart Wheat-Rice Production Technical Regulation, Climate-Smart Wheat-Corn Production Technical Regulation,* and *Climate-Smart Crop Production Measurement and Monitoring Method Regulation,* which have been included in the 2019 draft industry standards in the field of resources and environment. In the *Technical Guidelines for Green Agricultural Development (2018–2030),* the Ministry of Agriculture and Rural Affairs officially proposed to "develop climate-smart agriculture", which provides a policy basis for the popularization and application of CSA. We have gained rich experience and remarkable achievements in the innovation of technology training mode, popularization, and education of science popularization and application of new technology, as well as the transformation of production mode of new business entities.

It has been proved that CSA is an important mode of green agricultural development in China and can be regarded as a sustainable agricultural development mode integrated into the internationalization. CSA can effectively promote the development of green agriculture in China, realize intensive and high yield, promote cost savings and increase efficiency, bring about low carbon and environmental protection, institute environment-friendly measures, and increase farmers' income. It can become a leading model for the green development of agriculture in China and play an important role in ensuring national food security, resource and ecological security, improving the quality and efficiency of agriculture and rural revitalization. The development of CSA is conducive to China's extensive participation in global governance, implementation of international conventions on climate change, and substantive integration into the mainstream direction of international sustainable agricultural development.

The core of how to develop CSA as the mainstream mode of Chinese agriculture is to further promote the innovation of agricultural subsidy policy and technology mode. It is necessary to realize the development goals of prohibition, restriction, and promotion in accordance with the principles of "red, yellow and green light" and establish corresponding laws and regulations and subsidy policies as support. In particular, problems like the channels and varieties of agricultural subsidy policies, and the goals of each other are in conflict and should be solved. There is a need to establish production control standards and technology systems that are compatible with policies and management systems, and to radically change the research and promotion model that considers only crops, resources, environments, and individual technologies.

9.4 Development Proposals for CSA in China

9.4.1 *Introduce Climate-Smart Agriculture Development Plan and Supporting Incentive Policies and Measures As Soon as Possible*

Based on the policy system of foreign CSA development and China's agricultural production, social and economic status, the national CSA development strategy should be introduced as soon as possible, the overall target of carbon sequestration and emission reduction should be clarified, the priority areas of CSA should be put forward, and feasible implementation schedule should be formulated. In terms of supporting policies, a reasonable incentive and assessment system should be established, and financial subsidies and preferential policies should be given to farmers and agriculture-related enterprises that adhere to the concept of CSA.

9.4.2 *Ensure Stable and Sustained Financial Support from the State for CSA Technology Research and Development and Model Integration*

The government should strengthen multisectoral and multidisciplinary cooperation, increase investment in scientific research on new materials, new technologies and methods for CSA, strengthen research and development of theories and technologies, and accelerate the integration and application of energy-saving and emission-reduction models. Besides, it is worthy to establish a national long-term positioning trial network platform, form a unified data monitoring index system, measurement method rules and data collation standards, improve the data collection and release mechanism, and provide support for domestic and foreign research institutions to carry out relevant research.

9.4.3 *Strengthen the Integration, Demonstration and Application of Climate-Smart Agricultural Production Technologies*

(1) To strengthen agricultural meteorological prediction services and enhance the adaptability of agriculture to climate change.
The research of crop model and climate model started late in China, and the utility of agrometeorological forecast in the service formulation of agricultural production adaptation to climate change needs to be further strengthened. Strengthening the research on the impact of climate change on crop production is an important support to improve the adaptability of China's agricultural production to climate change.

(2) Vigorously develop climate adaptation technologies and models for disaster prevention and mitigation.
In the future, with the aggravation of climate change, the frequency of extreme abnormal weather, including responding to meteorological disasters such as low temperature, high temperature, drought, and flood, will increase. The key to deal with this change is to construct the technologies and models

including the selection of varieties, adjustment of sowing date, regulation of water and fertilizers, invasion of alien species, and the prevention and control of major crop diseases, pests, and weeds under the climate fluctuation.

(3) Optimize the planting mode.

Optimizing the cropping system and improving the productivity and sustainability of the farmland system play a significant role in addressing and mitigating climate change. This mainly includes the optimization of the planting mode of carbon sequestration and emission reduction, such as rice planting and cultivation, rotation, and fallow cultivation, intercropping, and multiple cropping. For example, the rational rotation mode of grain and soybean cropping can reduce the agricultural application of chemical fertilizers and improve the utilization efficiency of system resources. Through intercropping, the diversity of crop production system is enhanced and the system's ability to resist disasters is enhanced.

(4) Optimize soil carbon sequestration technology and model in farmland.

Promoting soil carbon sequestration is an important means of sequestration of carbon in the atmosphere to achieve climate change mitigation. It includes straw returning, conservation tillage, green fertilizer planting, organic fertilizer application, biochar, marsh fertilizer and other waste utilization, agroforestry planting system, non-point source pollution comprehensive prevention and control, ecological farmland construction and other technologies and models.

(5) Vigorously promote technologies and models for reducing greenhouse gas emissions in farmland.

Reducing greenhouse gas emissions from farmland ecosystems helps mitigate climate change. This mainly includes deep side fertilization, seed fertilizer co-planting, nitrification inhibitor or urease inhibitor addition, slow and controlled-release fertilizer, integrated water and fertilizer utilization, intermittent irrigation, rice dry direct seeding, facility seedling and other water-saving, as well as alternative pesticide emission reduction technologies and models for physical and biological control of diseases, pests, and weeds.

9.4.4 *Choose CSA Development Mode According to Local Conditions*

China has a vast territory and regional resource conditions varying between different areas of the country. The CSA development model should be rationally planned according to the regional agricultural development situation and existing problems. It should be supported by strengthening adaptation capacity and carbon sequestration so as to achieve a win-win situation in food security, farmers' livelihood improvement, and climate change mitigation. For example, in Northeast China, the carbon sequestration potential of farmland and resilience and adaptability of farmland to climate change should be enhanced. In the main rice-producing areas, greenhouse gas emission reduction should be the primary goal. In Northwest China and other ecologically fragile areas, it is necessary to improve the utilization efficiency of water and fertilizer resources and maintain the biodiversity of farmland. In pastoral areas, the focus should be on strengthening grassland ecological construction and improving the production efficiency of livestock products.

9.4.5 *Strengthen International Cooperation and Communications in Climate-Smart Agriculture*

The world today is faced with such global challenges and risks as climate change and food security. It is imperative for all countries to work together to cope with these challenges and risks. In recent years, China has made great achievements in carbon sequestration in agriculture, but there is still much room for emission reduction. It is necessary to continue to strengthen international communications and cooperation and learn from other countries, so as to provide a reference for dealing with global climate change. At the government side, in line with the principle of "mutual benefit, pragmatism and effectiveness" and "common but differentiated responsibilities", China should actively participate in and promote practical cooperation with governments and international institutions, and actively carry out cooperation with the Food and Agriculture Organization of the United Nations, the Global Environment Fund, the World Bank, and other institutions. Besides, China should promote the communications of policy and technical experience with the United States, the European Union, Australia, and other countries and regions, promote agricultural science and technology progress through joint tackling of key problems, and share the achievements of CSA.

Index

Printed in the United States
by Baker & Taylor Publisher Services